ROCKWELL INTERNATIONAL SPACE SHUTTLE

NASA

OV-101 "Enterprise" in flight over Edwards AFB, California during the course of its second Approach and Landing Test (ALT) on September 13, 1977. It had been carried to altitude and launched from NASA's Boeing 747 shuttle carrier aircraft. The flight, with Joe Engle as commander and Richard Truly as pilot, lasted 5 min. 28 sec.

This publication is dedicated to the memories of ten United States Astronauts who have perished in flight and flight preparation accidents. It is a dangerous profession, and those of us on the ground try our hardest to make it as safe as possible. Sometimes we don't succeed. Godspeed: *Apollo 1* (AS-204); *STS-33* (51-L/*Challenger*).

CREDITS:

The following people were invaluable in the assistance they rendered in the preparation of this book: Wesley B. Henry at the Air Force Museum; the KSC Technical Library Documents Section staff (William Cooper, Dorothy Price, Jane Page and Donna Atkins); Lisa Vasquez-Morrison at JSC; Ralph Esposito at KSC; Lee Browndorf at VLS; Phil Green at Rockwell International; Amos Chrisp at MSFC (for his efforts to save *Pathfinder*); Gerald Balzer; Walt Boyne; Richard Hallion, Ph.D; Gayle Lawson; Paul Minert; Mick Roth; Erik Simonsen; Doug Slowiak; Mike Wagnon; Barbara Wasson; and Jay and Susan Miller.

BACKGROUND:

Reusable spacecraft have been discussed for over sixty years; Robert Goddard and Konstantin Eduardovich Tsiolkovskiy wrote of them during the 1920s. In Germany, others put form as well as thought to the idea of winged rockets. A Viennese scientist, Max Valier, believed rocket engines would replace aircraft internal combustion engines and lead by natural evolution to winged spacecraft that would fly back and forth to space. Working in the mountains of western Germany, Valier produced small glide-rockets, eventually working up to a 30 lb., seven foot long piloted vehicle named *Stork*. This glide-rocket used a 165 lb. th. engine but produced dubious results, so during 1929, Fritz von Opel, who had funded many of Valier's experiments, produced a larger, manned glide-rocket that flew for about 10 minutes at 100 mph. This 600 lb. vehicle crashed upon landing, and the ensuing fire almost cost von Opel his life. This probably was the first manned rocket flight, and it accurately forecast the difficult road that lay ahead in attempting to send men into space.

SÄNGER AND BREDT

The greatest contributions to early studies were undoubtedly made by one man, Eugene Sänger, who conceptualized the development of spacecraft capable of flying into space and returning to normal aircraft type landings while he was a doctoral candidate at the Viennese Polytechnic Institute during 1929. Out of his studies emerged a 1933 concept dubbed *Silbervogel* (Silverbird), a winged aircraft propelled by a liquid fuel rocket engine burning kerosene and liquid oxygen, capable of reaching Mach 10 and altitudes in excess of 100 miles. This first *Silverbird* had a "spindle-shaped" (Sänger's words) fuselage and straight wings of low-aspect ratio having sharp leading edges with wedge airfoil sections. Sänger estimated during 1934 that with a lift-to-drag (L/D) ratio of 5.0, the craft could obtain Mach 13 speeds at the moment of fuel exhaustion followed by a deceleration to a steady Mach 3.3 cruise. An operating altitude of 160,000 ft. and a range of over 3,100 miles were projected. In collaboration with mathematician Irene Bredt, whom he later would marry, Sänger would continue to refine this design for the next ten years.

Towards the end of the 1930s, the *Silverbird* concept had developed into a flat-bottom "laundry-iron" shape having low-aspect ratio wings with wedge airfoils, and a similar horizontal stabilizer with twin endplate vertical tails. The Sänger-Bredt *Silverbird* was theoretically capable of lifting a four-ton payload into low-earth orbit using a 200,000 lb. th. liquid fuel rocket engine which was ignited in flight after launch by a Mach 1.5 rocket sled. Following deployment of its payload, the *Silverbird* would return to earth in a series of semi-ballistic skips, each skip becoming shallower and shorter than the previous, until finally gliding to a conventional runway landing. Alternately, the design could follow a suborbital trajectory and deliver an eight-ton payload to an antipodal point halfway around the world from its launch site using the skip-glide technique. Skip-gliding would lose favor during the 1960s when further study showed it would expose the airframe to prolonged and extreme heating.

During the Second World War, Sänger and Bredt concentrated on their work on the development of more con-

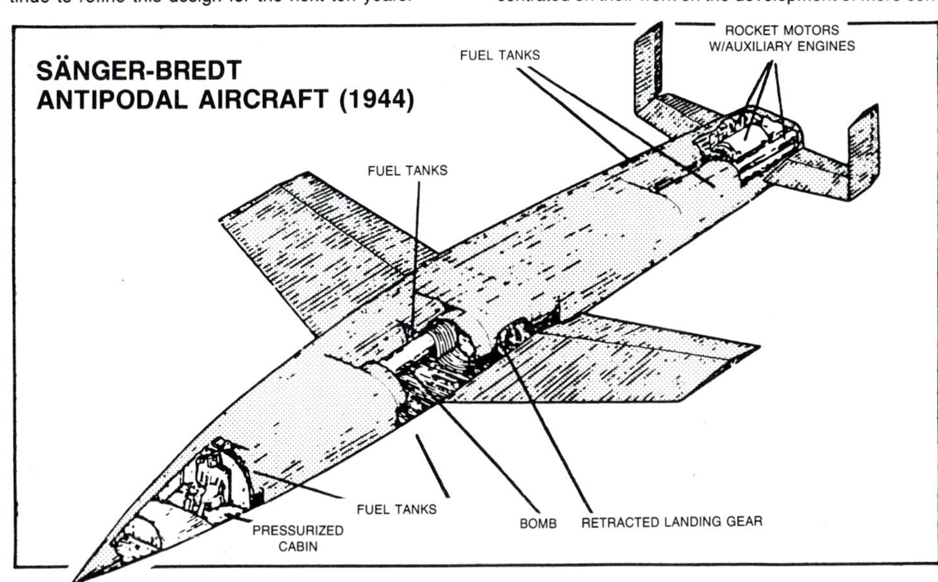

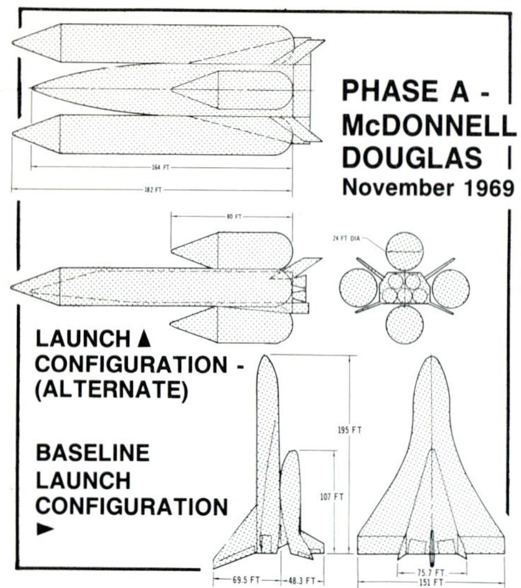

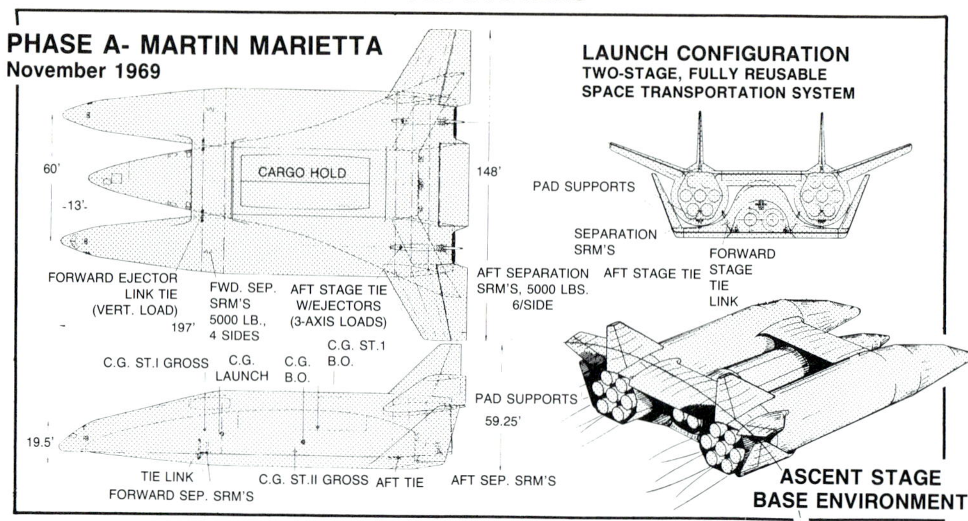

ventional high-speed aircraft and propulsion systems. Nevertheless, they found time to continue study of the *Silverbird* by advocating it as an "Amerika Bomber" to the Luftwaffe. While little official support was forthcoming, the pair did publish a report during 1944 under the title *Concerning Rocket Propulsion for Long-Range Bombers*, and several copies of this subsequently were captured by American and Soviet technical intelligence teams. Reportedly, Soviet Premier Josef Stalin was so impressed by the report that he sent a team into Western Europe to locate and kidnap Sänger and his wife. The plan was fouled by the French Secret Service, and the pair continued their work, first in France and subsequently in West Germany. During 1964, shortly before his death, Eugen Sänger was at work as a consultant to Junkers, developing a delta-wing craft (the RT-8-01) bearing a close resemblance to his work of the previous three decades. This design continued for several years as the *Sänger-I* before being dropped for a lack of funding and official interest. During 1987, a European group led by Messerschmitt-Bölkow-Blohm (MBB) announced plans for the development of the *Sänger-II* two-stage-to-orbit reusable space transportation system.

EARLY EXPERIMENTS

Though much of the early work on reusable spacecraft followed the inspiration of the Sänger-Bredt *Silverbird*, many German rocket engineers brought to the U.S. under the auspices of *Project Paperclip* already had participated in a major attempt to develop and fly a lifting reentry craft. In the midst of WWII, engineers working on the development of the A-4 (V2) rocket at Pëenemunde had examined a series of concepts for multi-stage vehicles utilizing winged upper stages. One notable effort was the A-9/A-10 combination, which planners envisioned as a large booster (the A-10) topped by a winged second stage (A-9) capable of delivering a one ton warhead over a range of 3,000 miles. The A-10 would boost the A-9 into the upper atmosphere, and then fall away to its destruction. The A-9 would fire its engine, continue down-range in a ballistic arc, then transition to a terminal glide at Mach 3.5 towards its target. A larger, orbital version of the winged A-9, launched by a two-stage booster known as the A-11/A-12 and capable of lifting 30 tons to an orbiting space station was under study at the war's end. In a separate, but related effort, Ludwig Roth supervised the design and building of two swept-wing derivatives of the A-4. The first of these, designated A-4b crashed and exploded shortly after launch on January 8, 1945. The second, launched on January 24, reportedly exited the atmosphere, completed a stable, ballistic reentry, and began a Mach 4 glide to earth. During the glide one of the wings separated from the vehicle, probably from unexpectedly high flight loads, and the A-4b broke up. German winged entry research efforts were terminated as the war in Europe came to an end, and neither the Americans, nor the Soviets seemed interested in continuing it.

During 1949, California Institute of Technology professor Hsue-Shen Tsein (who would later return to the Peoples' Republic of China and lead the PRC's development of ballistic missiles and space launch systems) concluded that a sufficient technological base existed for designers to develop a Mach 12 transcontinental transport powered by rocket engines burning liquid fluorine and liquid hydrogen. Two years later, at a space symposium held at the Hayden Planetarium, enthusiasts discussed the future of spaceflight, and Wernher von Braun, by now working for the Army on ballistic missile research, advanced the concept of a three-stage launch vehicle culminating in a huge winged spacecraft powered by five engines burning nitric acid and hydrazine. Theoretically capable of placing 36 tons into orbit, the craft was to be built of steel, have a length of 77 ft., and span 156 ft. In an elaboration of this theory in *Collier's* the next year, von Braun predicted the eventual development of space stations serviced by reusable winged spacecraft. While none of these ambitious concepts resulted in specific development programs, they did offer an opportunity for engineers to evaluate the current state-of-the-art in reentry and propulsion technologies, as well as to fuel the imagination of the general public and science fiction writers.

THE X-PLANES

During the 1950s and 1960s, the U.S. Government sponsored a series of experimental research aircraft commonly known as the "X-planes". The early X-planes, such as the Bell X-1 and Douglas D-558-1, gave aviation its first experience with supersonic flight. By 1956, these craft had achieved speeds in excess of Mach 3, and had flown to altitudes above 126,000 ft. The X-planes were the first vehicles to encounter the kinds of control difficulties that would demand the development of thruster reaction controls, and require complex alloy structures to withstand the temperatures of high speed flight. Other benefits from the early X-planes included insight into the problems of inertial coupling, exhaust impingement upon control surfaces, and an appreciation for the complex physiological protection necessary for crewmen. (See *The X-Planes*, by Jay Miller, Aerofax Inc., 1988)

During the early 1950s, L. Robert Carman and Hubert M. Drake, two engineers at the NACA High-Speed Flight Station (now the Ames-Dryden Flight Research Facility) advocated an ambitious proposal for a small hypersonic research vehicle. The Drake/Carman report, released on May 21, 1952, predicted that a 100,000 lb. (gross) vehicle using water-alcohol and liquid oxygen propellants could obtain Mach 6.4 at 660,000 ft. Using this vehicle as a carrier aircraft, a vehicle the size and weight of the Bell X-2 could be launched at Mach 3 and 150,000 ft., attaining speeds up to Mach 10 at altitudes approaching 1,000,000 ft. This general concept of a two-stage vehicle with both the orbiter and launch vehicle being reusable would shape most early space shuttle studies.

On September 8, 1952 the NACA formed a hypersonic research study group chaired by Clinton E. Brown, and this body made a number of recommendations for future NACA high-speed research projects. Although they realized significant problems remained to be solved, the committee was optimistic that the time was not too far off when hypersonic boost-glide vehicles could be developed. After reviewing the Drake/Carman report, on June 23, 1953 the group endorsed a proposal by David Stone of the Piloted Aircraft Research Division involving boosting a modified Bell X-2 to Mach 4.5 and 300,000 ft. using two Sargent solid rockets. This project did not reach the hardware stage.

BOMI, ROBO AND *BRASS BELL*

Following the Second World War, Walter Dornberger, former director of the Pëenemunde center, had emigrated to the U.S. to join the staff of Bell Aircraft. While at Bell, Dornberger advocated the development of a Sänger-like flat-bottom lifting reentry spacecraft, and during 1952, even journeyed to France in a vain attempt to persuade Sänger and his wife to come to the U.S. to join Bell.

Early Bell designs were variations of the Sänger-Bredt *Silverbird*, but the wedge profile straight wings soon gave way to a delta planform more typical of later concepts. The company devoted a great deal of attention to the probems of reentry, particularly in the field of thermal protection, where Bell explored both active (such as liquid circulation) and passive (radiative and heat-sink) schemes. During 1952 Bell proposed building a piloted boost-glide bomber-missile dubbed *Bomi* to the AF. *Bomi* was a two-stage vehicle featuring a five engine, delta wing booster, and a smaller three engine double-delta winged glide-rocket. The pair would use the booster's engines for two minutes, then the glide-rocket would separate and ignite its engines. The booster would then glide back to its base for a runway landing and reuse. During the period of maximum aerodynamic loading ("max-q"), the glide-rocket would throttle down two of the engines to ease air-loads. The lower stage was 100 to 120 ft. in length with a wing span of 60 ft. and carried a crew of two. Most of the structure was to be aluminum, with titanium leading edges. The upper stage was 60 ft. in length with a 35 ft. wing span and carried a single pilot and a payload consisting of a single 4,000 lb. nuclear bomb. It was to be built entirely of titanium and use a radiant cooling system. A gross lift-off weight of 600,000 to 800,000 lbs. was projected. The propellants for both stages were un-dimethylhydrazine (UDMH) and nitrogen tetroxide. *Bomi* technically was a suborbital design having a 3,000 mi. range, and capable of Mach 4 at 100,000 ft. An orbital version of *Bomi* also was proposed. This version would have a lower stage 144 ft. long built entirely of titanium. The new upper stage would be 75 ft. long, and carry a payload of 14,000 lbs. It was covered with a graphite epoxy ablative heat shield on a honeycomb backing. It was hoped that the graphite-epoxy could be sprayed-on after a mission to renew the heat shield, but this was one of the more serious unknowns in the vehicle's development. Two nuclear bombs could be carried in A-5 *Vigilante*-style rear-ejecting bomb-bays. The propellants for the orbital-version were changed to liquid oxygen and liquid hydrogen. The AF evaluation of *Bomi* pointed out several conceptual flaws, but on April 1, 1954 the service awarded Bell a contract to study Weapons System MX-2276 based on the boost-glide concept. MX-2276 was to perform both reconnaissance and bombardment roles, and was to be capable of 15,000 mph at 259,000 ft. with a projected range of 10,600 miles. By December 1, 1955, $420,000 had been expended by Bell on *Bomi*/MX-2276 studies. Although unknown at the time, *Bomi* would have a significant influence on future space shuttle concepts.

A year earlier, on January 4, 1955, Bell participated in an investigation applying the technologies being developed for *Bomi* to a reconnaissance aircraft known as System 118P. Bell's design for System 118P included a two-stage rocket to boost a glider to 165,000 ft. at a

INTEGRAL LAUNCH AND REENTRY VEHICLE SYSTEMS

PHASE A - GENERAL DYNAMICS/CONVAIR October 1969

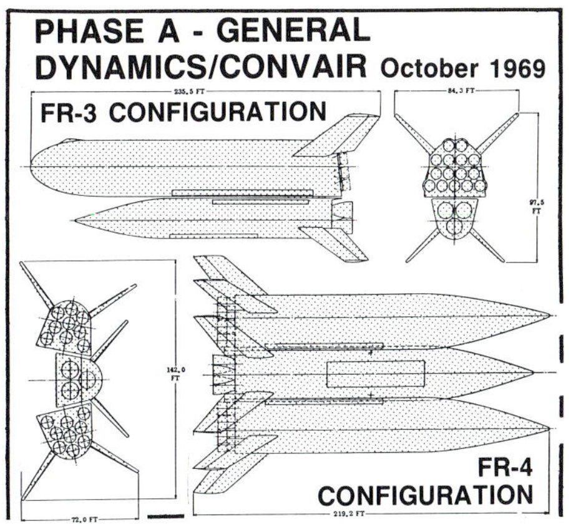

PHASE A - NORTH AMERICAN ROCKWELL December 1969

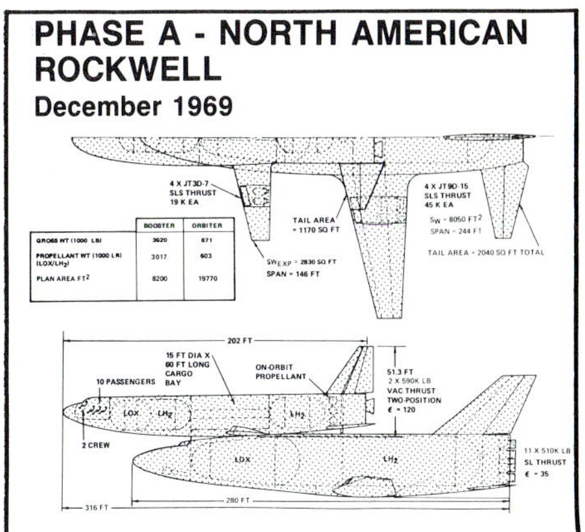

velocity of Mach 15. The *Bomi* and System 118P efforts were combined on March 20, 1956 and the AF awarded Bell a $746,000 (later raised to $1.2 million) contract for Reconnaissance System 459L, also known as *Brass Bell*. By December 1956, Bell had defined a manned, two-stage system which would be propelled by *Atlas* ICBM engines over 5,500 miles and a velocity of 12,000 mph at an altitude of 170,000 ft. Bell engineers reasoned that with the addition of another booster stage, the range of *Brass Bell* could be extended to 10,000 miles with a maximum speed of 15,000 mph.

During June 1956 the AF had released System Requirement 126 for the study of a rocket-bomber later designated *Robo*. The purpose of the study was to determine the feasibility of a manned, hypersonic bombardment system, and one of the technologies to be investigated was a glide-rocket similar to *Bomi/Brass Bell*. In support of *Robo* and *Brass Bell*, the AF initiated a research program titled "Hypersonic Weapons Research and Development Support System" (*Hywards*). Designed to provide research data on aerodynamic, structural, human factor and component problems associated with high-speed (Mach 15-18) atmospheric flight and reentry, *Hywards* also was to serve as a test craft for the development of subsystems to be employed in future boost-glide systems.

During 1957, Dr. Krafft A. Ehricke and Dornberger collaborated to design a two-stage passenger version of *Bomi*. The stages were to be mounted in piggy-back fashion, with the lower stage having five rocket engines, and the passenger stage having three. Each stage was of a delta planform, and it was anticipated that the vehicle would take-off with both stages firing until 130 seconds after launch when the lower stage would separate and glide back to land. The passenger stage would continue, completing a 3,000 mi. flight in about 75 minutes at an altitude of 150,000 ft.

On October 4, 1957 the Soviet Union orbited *Sputnik* the first artifical earth satellite. The shock to the West was profound, and less than a week later, the AF consolidated *Robo*, *Brass Bell* and *Hywards* into a single development program called *Dyna-Soar* (*Dynamic Soar*ing—as Sänger had called skipping reentry). A conference opened on October 15, 1957 at NACA's Ames Laboratory to determine the direction of *Dyna-Soar*, and three different approaches to manned spaceflight were proposed. A minority, led by Maxime Faget, argued for a purely ballistic Allen-type blunt reentry shape—essentially what emerged as the capsule for *Project Mercury* (Interestingly, later NASA capsule designs had a substantial increase in lift-to-drag ratio. In fact, *Apollo*, with a L/D of 0.8 rivaled some of the early lifting-bodies in cross-range capability, though it was never used.). Another minority favored the "lifting-body" approach of tailoring the design of a blunt reentry shape to provide a modest L/D permitting limited maneuvering during the reentry profile. The remainder of the conference endorsed the concept of a flat-bottom hypersonic glider. On December 21, 1957 the AF Research and Development Command issued System Development Directive 464L for the first phase of *Dyna-Soar*, a single-seat hypersonic boost-glide technology demonstrator.

The subsequent design competition to develop the *Dyna-Soar* vehicle consumed the better part of two years and involved nine contractor teams. Of these, only Boeing and a combined Martin-Bell team actually attempted design of a true orbital spacecraft; the others envisioned developing some form of hypersonic research craft that could eventually spawn an orbital vehicle. Not content to pursue this intermediate approach, the AF directed Boeing and Martin-Bell to undertake more detailed studies, and as a result, Boeing was declared the winner on November 9, 1959. Martin received a contracted to develop the booster, a variation of the company's *Titan* ICBM.

The craft eventually emerged as a slender delta planform with a rounded and slightly upward tilting nose and twin endplate vertical fins. Construction materials consisted primarily of a René-41 superalloy for the structure, a columbium superalloy heat shield, and a graphite/zirconia composite nose cap. To demonstrate the flying characteristics of the vehicle through Mach 2, *Dyna-Soar* was scheduled for twenty air drop tests from a B-52 beginning during July 1963. Starting during November 1963, five unmanned flights were to be launched from Cape Canaveral, Florida, to landing sites in Mayaguana in the Bahama Islands, and Fortaleza, Brazil, with velocities ranging from 6,000 mph to 13,000 mph. Eleven manned flights would leave Cape Canaveral starting during November 1964, for landings in Santa Lucia in the Leeward Islands, and Fortaleza. These suborbital flight tests were to cost $493.6 million, including the development costs of the vehicle. During April 1960, Professor C. D. Perkins (Assistant Secretary of the AF for Research and Development) requested the *Dyna-Soar* office to examine the use of a *Titan II* booster for the suborbital test flight instead of the planned *Titan I*. It was anticipated the performance of the *Titan I* would be inadequate for the rapidly increasing weight of *Dyna-Soar*, and modifications to the glider should be minimal since a modified *Titan II* with a *Centaur* upper stage already was being considered for the later orbital missions. On April 26, 1961 the *Dyna-Soar* office elaborated on the previous flight test plan: following twenty air launched tests started during January 1964, the first of two unmanned ground launched tests would occur during August 1964, with the first of 12 manned suborbital flights during April 1965 on top of a *Titan II*.

The April 26 plan also detailed follow-on tests, including one-orbit flights from Cape Canaveral to Edwards AFB, California starting during April 1966. An interim operational vehicle, capable of fulfilling reconnaissance, satellite inspection, space logistics and bombardment missions was anticipated by October 1967. The complete weapons system, including space-to-earth and space-to-space missiles could be available by late 1971.

However, the newly established AF Space Systems Division (SSD) announced a manned satellite inspector program, SAINT II, on May 19, 1961. The proposed SAINT II demonstrator was a two-man lifting-body derived from the SAINT I unmanned prototype then under development by SSD. The vehicle was to be launched by a *Titan II* with a new fluorine/hydrazine powered upper stage called *Chariot*. Twelve orbital missions were scheduled, with the first unmanned flight occurring early during 1964, and the initial manned launches set for later that year. SAINT II was to have both a low-earth (200-300 mile) and a far-earth (1,000 mile) capability. Estimated costs were $413.8 million during 1962 through 1965, and SAINT II was considered a contender for *Dyna-Soar* funds.

On October 7, 1961 *Dyna-Soar* officials unveiled a plan to restructure *Dyna-Soar* again, this time to include the development of a far-earth demonstration vehicle in addition to the low-earth vehicle originally envisioned. The plan eliminated the suborbital test phase, and reduced the air launched program to fifteen flights. The program office anticipated the first unmanned orbital flight during November 1964, and the first piloted flight during May 1965 on top of a *Titan III* (essentially a *Titan II* with two strap-on solid fuel rocket boosters). The next five flights would be piloted multi-orbital missions. The ninth test flight, occurring during June 1966, would be an unmanned exploration of the velocities needed for far-earth missions and the remaining nine flights would be piloted with the purpose of demonstrating military missions such as satellite inspection and reconnaissance. The flight test program was to terminate during December 1967 with a total program cost of $921 million, but no specific production schedules were included in the revised plan.

Secretary of Defense McNamara endorsed the restructuring on February 23, 1962. *Dyna-Soar* was now a research and development program for a military system to explore and demonstrate maneuverable reentry of a piloted orbital glider which could execute runway landings at some preselected site. After considering several designations (including XMS-1 for "Experimental Manned Spacecraft - One"), the *Dyna-Soar* officially was designated X-20 on June 19, 1962.

Although by this time the SAINT II effort had succumbed to definition and funding problems, *Dyna-Soar* found itself with a new competitor—*Blue-Gemini*—later to evolve into *Gemini-B*. On January 18, 1963, Secretary McNamara directed a comparison study between the X-20 and NASA's *Gemini* to determine which represented the more feasible approach to a military capability. The

INTEGRAL LAUNCH AND REENTRY VEHICLE SYSTEM PHASE B - McDONNELL DOUGLAS/MARTIN MARIETTA December 1970

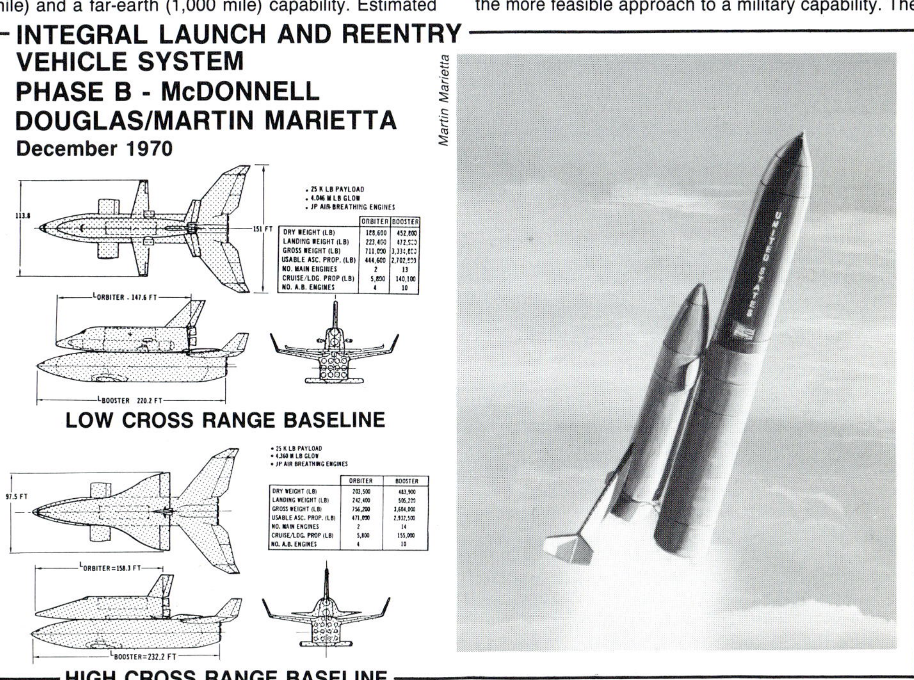

MSC 'DC-3' CONCEPT
April 1970

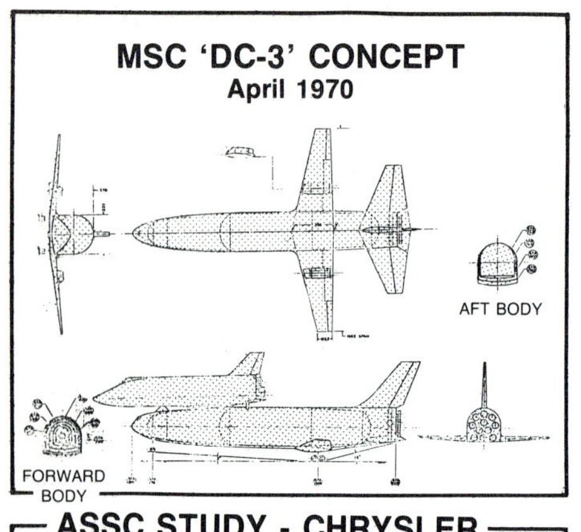

ASSC STUDY - CHRYSLER
June 1971

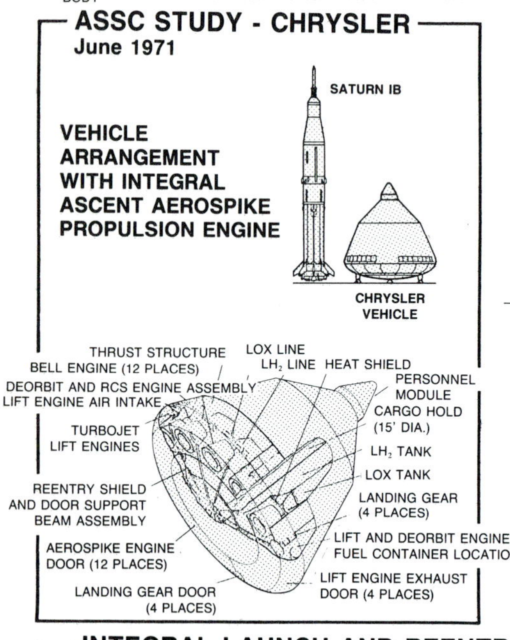

ASSC STUDY - GRUMMAN
July 1971

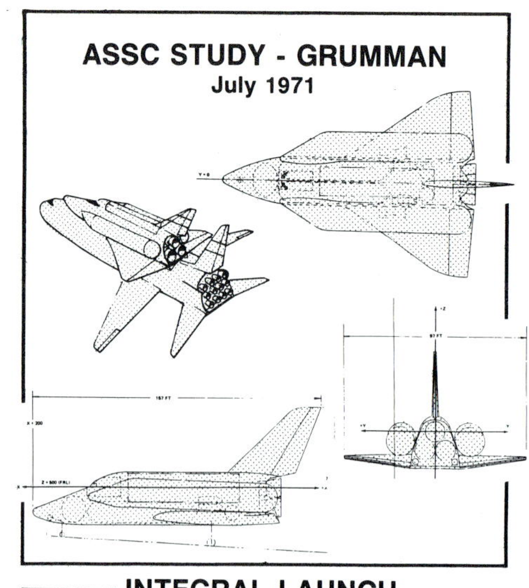

INTEGRAL LAUNCH AND REENTRY VEHICLE SYSTEM PHASE B - ROCKWELL
February 1971

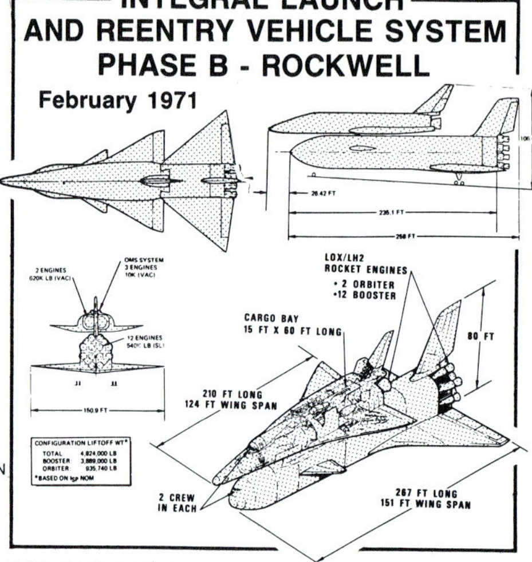

INTEGRAL LAUNCH AND REENTRY VEHICLE SYSTEM PHASE A CONCEPTS McDONNELL DOUGLAS

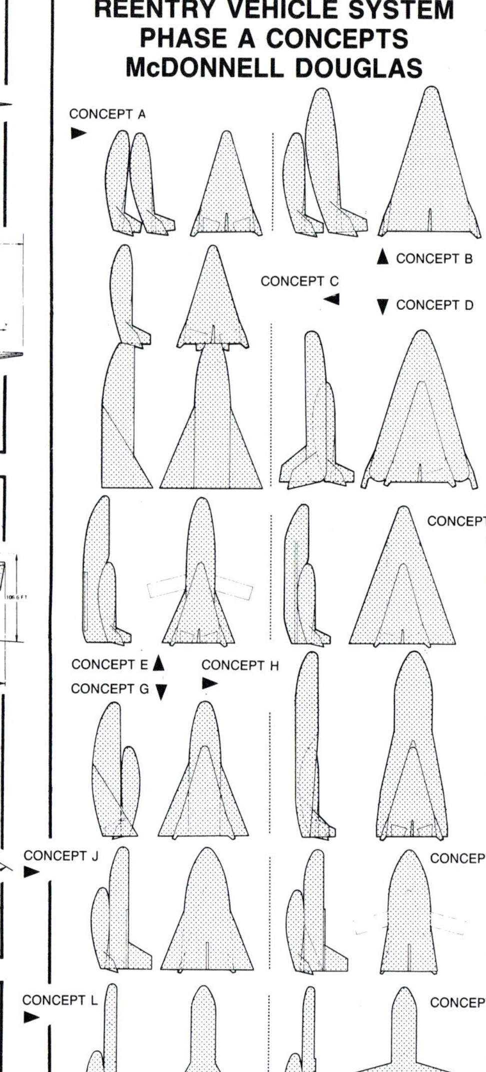

INTEGRAL LAUNCH AND REENTRY VEHICLE SYSTEM EARLY MSC-040C-1 CONCEPT
March 1972

CREW SPACES
MAIN ENGINE
53 FT.
120 FT.
AIRBREATHING ENGINE
MAIN GEAR
80 FT.
PAYLOAD BAY

INTEGRAL LAUNCH AND REENTRY VEHICLE SYSTEM PHASE A - LOCKHEED

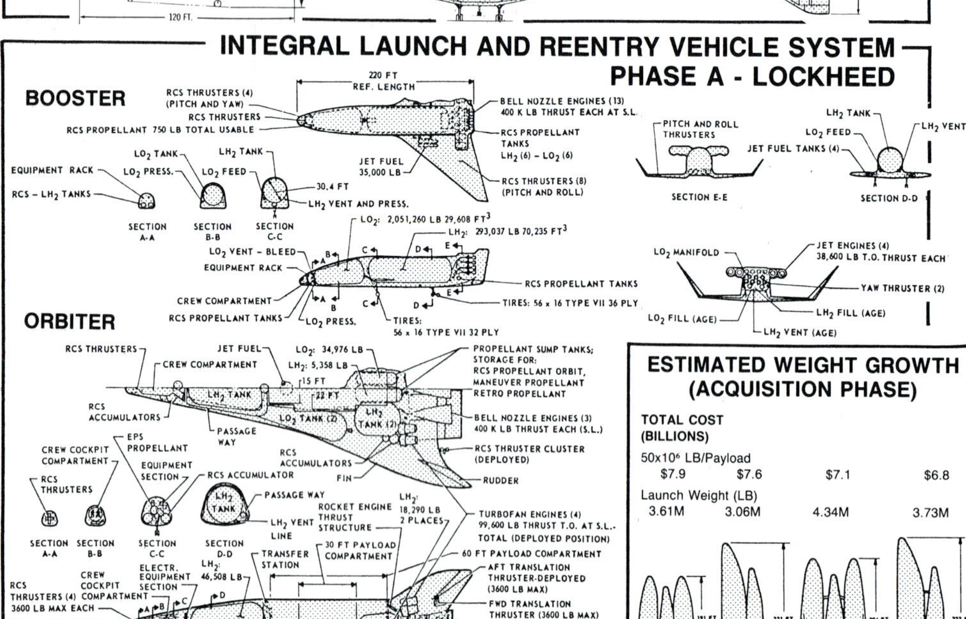

ESTIMATED WEIGHT GROWTH (ACQUISITION PHASE)

TOTAL COST (BILLIONS)				
50x10⁶ LB/Payload	$7.9	$7.6	$7.1	$6.8
Launch Weight (LB)	3.61M	3.06M	4.34M	3.73M
	TRIAMESE	TWO-STAGE	TRIAMESE	TWO-STAGE

ASSC STUDY - LOCKHEED
BASELINE CONFIGURATION

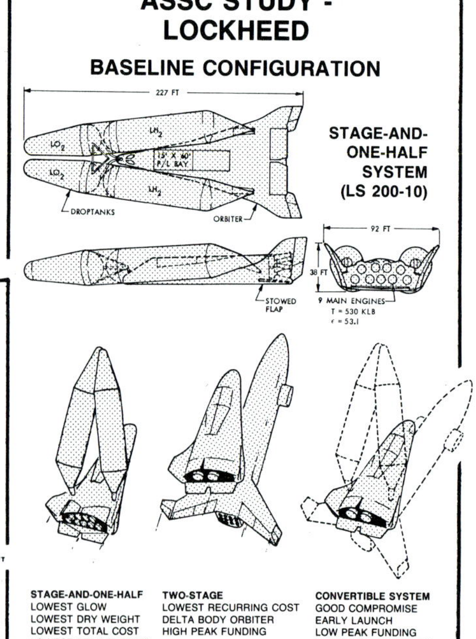

STAGE-AND-ONE-HALF
LOWEST GLOW
LOWEST DRY WEIGHT
LOWEST TOTAL COST
LOWEST PEAK FUNDING
LOWEST RISK

TWO-STAGE
LOWEST RECURRING COST
DELTA BODY ORBITER
HIGH PEAK FUNDING

CONVERTIBLE SYSTEM
GOOD COMPROMISE
EARLY LAUNCH
LOW PEAK FUNDING
LOW RECURRING COST
HIGHEST TOTAL COST

AF Chief of Staff, Gen. Curtis LeMay, insisted the purpose of AF participation in the *Gemini* program was limited to obtaining experience and information concerning manned space flight, and *Dyna-Soar* was the preferred operational system. Secretary McNamara, however, was not convinced the AF could demonstrate a pressing need for any manned orbital capability.

Meanwhile, *Dyna-Soar* was again having to restructure its flight test program to meet funding reductions. While final budget figures still were pending for FY64, the impact of the 1961 redirection of the *Dyna-Soar* program was becoming apparent. The first *Dyna-Soar* development plan had definitive military objectives leading to the development of orbital reconnaissance and bombardment vehicles. These goals had been altered, and the major emphasis placed on the development of suborbital and orbital research vehicles. In spite of intensive comparative studies with the manned SAINT II and *Blue-Gemini* concepts, the X-20 officially was described as an experimental research vehicle. By mid-1963, DoD was seriously questioning the necessity for the *Dyna-Soar* program and it appeared that the alternatives for the X-20 had been severely narrowed: direct the program towards achieving military goals, or terminate it.

A May 1963 study by the AF Space Systems Division and the Aerospace Corporation concluded the X-20 could be adapted for testing of military subsystems and operations. *Dyna-Soar* had a payload volume of 75 cubic ft., sufficient power, and enough cooling capacity to accommodate military subsystems. Since deceleration occurred slowly during lifting reentry, it would provide a safe physiological environment for the transfer of personnel and equipment from space stations. Lastly, significant information for the development of future maneuvering reentry spacecraft could be obtained from the X-20 program. Total cost to develop and test a militarized X-20A would be $228 million, in addition to the basic X-20 program costs.

The *Dyna-Soar* program office also completed a study concerning the use of the X-20 for anti-satellite missions. A proposed X-20B configuration would have an interim operational capability of satellite inspection and negation. Two flights would be added to the X-20 program to demonstrate the operational capability at an additional cost of $227 million. To conduct a 50 flight operational program would cost $1.2 billion during fiscal 1956-1972. A satellite inspection version, the X-20X, had provisions for a two-man crew with a 14-day endurance, and could inspect targets as high as 1,000 miles. The first flight of the X-20X version was projected for September 1967 and would need $350 million in additional funding.

Although military missions for the *Dyna-Soar* were identified, convincing Washington that they were valid proved more difficult. A military presence in space could be more rapidly, and much more economically achieved by participating in NASA's *Gemini* program. When, during October 1963, Deputy Defense Secretary Harold Brown recommended a permanently manned military space station serviced by modified *Gemini* capsules, the X-20 had been dealt a death blow.

On December 10, 1963 Defense Secretary McNamara cancelled *Dyna-Soar* in favor of model testing via the ASSET program, and reappropriated the X-20 funding to the Manned Orbiting Laboratory (MOL). Thus ended the first serious attempt to build a reusable manned spacecraft. At the time of its cancellation, *Dyna-Soar* was 2-3 years from first flight; $410 million already had been spent on its development and another $373 million was expected to be spent prior to first flight. Even if it had never flown an orbital mission, it would have provided valuable information on reentry flight control and heating problems, something that was seriously lacking during development of the space shuttle ten years later.

During April 1961 the McDonnell Aircraft Corporation had signed a cost-plus fixed fee contract with the AF Flight Dynamics Laboratory for the construction of six experimental ASSET (Aerothermodynamic/elastic Structural Systems Environment Tests) reentry gliders that roughly resembled the now cancelled *Dyna-Soar*. The gliders were 5 ft. 8.5 in. long and weighed in at 1,100 to 1,200 lbs. The ASSET program consisted of two efforts: the Aerothermodynamic Structural Vehicles (ASV), and the Aerothermoelastic Vehicles (AEV); McDonnell would build four of the former and two of the latter. All had a sharply swept (70°) low aspect-ratio delta wing spanning 4 ft., 7 in. and providing 14 square ft. of area. A flight control system featuring hydrogen peroxide reaction thrusters maintained flight attitude after launch on top of *Thor* (AEV) or *Thor-Delta* (ASV) boosters. (These *Thor* IRBMs were part of a group returned from the United Kingdom after having stood nuclear alert for several years and were available to the research project at extremely reasonable costs.) Although superficially similar, and sharing common subsystems, the two craft differed completely in mission and research capabilities and this was reflected in totally different flight profiles. The ASVs had a mission to determine temperature, heat flux, pressure distribution and evaluation of materials and structural concepts during hypersonic gliding reentry. These were boosted to altitudes of 190,000 to 225,000 ft., and velocities of from 16,000 to 19,500 ft. per second, giving them a range varying from 1,000 to 2,300 miles. On the other hand, the AEVs would be boosted to 168,000 and 187,000 ft. at a 13,000 fps velocity, obtaining ranges of 620 and 830 miles. Their mission was to study hypersonic panel flutter phenomenon, and the effect of hypersonic flight on aerodynamic control surfaces. Although the second vehicle (an ASV) was destroyed as a result of booster failure, and several vehicles were not recovered from the Atlantic off the Eastern Test Range, the program was considered highly successful, and demonstrated that winged reentry vehicles could successfully transverse the upper atmosphere. The total ASSET program costs, including boosters, was $41 million.

The ASSET program furnished useful information on the aerothermodynamics and aeroelastic characteristics on a generic winged shape reentering the atmosphere at near-orbital velocities. During 1966-67 the AF would fly a totally different kind of reentry vehicle: an ablative-cooled lifting-body built by the Martin Company. Unlike ASSET, which had sacrificed an optimum aerodynamic configuration in favor of a large internal volume (to contain instrumentation), the PRIME (Precision Recovery Including Maneuvering Entry) shape emphasized aerodynamics with a carefully derived external shape. Again, unlike the structures and heating research oriented ASSET, PRIME explored the problems of maneuvering reentry, including pronounced cross-range maneuvers up to 710 miles off the ballistic track. A secondary objective was to acquire design data and to develop configuration characteristics pertinent to possible future manned reentry vehicles. It also generated a low-speed piloted demonstrator known as PILOT (Piloted Lowspeed Tests). ASSET, PRIME and PILOT, in addition to numerous ground test programs including the classified wind tunnel project M-103, were collectively known as START (Spacecraft Technology and Advanced Reentry Tests), and officially were designated USAF Program 680A.

As early as November 1960, the Martin Company had undertaken study of a variation of the Ames M1 lifting-body shape in support of the AF's SAMOS film return surveillance satellite. Engineers under the direction of Hans Multhopp eventually rejected the M1 shape, opting instead for the A3 configuration of the Aerospace Corporation, but modifying it to a more streamlined variation known as the A3-4, or SV-5. The AF selected the SV-5D, a subtle variation of the basic SV-5 shape, for the PRIME spacecraft. At this point it was planned that four PRIME vehicles would be launched on *Atlas* boosters from Cape Canaveral, and recovered after 3/4 of an orbit near Hawaii. Each of the planned PRIME shapes was 6 ft. 8 in. long, 25 in. high, spanned 3 ft. 10 in. and had a 34 in. cross-section. A hypersonic L/D ratio of 1.3 was expected. During flight, approximately 240 aerodynamic, thermal, guidance and system diagnostic measurements would be returned to the ground via VHF telemetry.

As eventually built, the SV-5D (later designated X-23A) was an 890 lb. aluminum lifting-body with an ablative heat shield and a carbon-phenolic nose cap. Since the PRIME test flights would terminate at Mach 2.0 with the deployment of a drogue ballute, adequate stability and control could be maintained with a movable lower flap and fixed upper flaps and rudders. The fixed, deflected upper flaps provided aerodynamic trim at high angles of attack, and the fixed, deflected rudder maintained directional stability during hypersonic and supersonic flight. The capability existed to make the fixed control surfaces operational since it was expected that an operational data return variant would make a horizontal runway landing instead of being recovered in the air. A hypersonic L/D ratio of 1.0 was achieved. The first PRIME vehicle was launched from Space Launch Complex 3 East (SLC-3E) at Vandenberg AFB on December 21, 1966 on a trajectory simulating reentry from low-earth orbit with a zero cross-range maneuver. The performance of the *Atlas* and SV-5D was superb through the boost and reentry phases resulting in a ballute deployment 4,300 miles downrange and within 900 ft. of the preselected target point. The ballute deployed at 99,850 ft., and when the SV-5D had descended to a little less than 45,000 ft., the main recovery parachute began to deploy but did not complete its extraction sequence. The vehicle fell into the Pacific and was lost. However, all flight objectives (except recovery) had been met, and over 90% of all possible telemetry had been successfully received. The second PRIME flight occurred on March 5, 1967, and successfully completed a 654 mile cross-range maneuver—the first ever for a returning spacecraft. Due to a failure in the parachute separation process (several stringers had failed to be cut, resulting in the vehicle being suspended in a manner the recovery of JC-130B could not snag) the vehicle again was lost in the Pacific. On April 19 the third SV-5D was launched in a trajectory simulating reentry from low-earth orbit with a maximum (710 mile) cross-range maneuver. Performance of the vehicle and all of its subsystems was perfect, and this time everything worked well for recovery. A waiting Lockheed JC-130B successfully snagged the SV-5D at 12,000 ft. less than five miles from its preselected recovery site. After a complete inspection by a Martin/USAF team, this third vehicle was found to be in satisfactory shape to be used again if needed. This was an important milestone in demonstrating the reusability of lifting reentry spacecraft. Satisfied with the results of the first three flights, and facing a funding cutoff, the AF cancelled the last PRIME flight, and the two surviving vehicles are on exhibit at the AF Museum at Wright-Patterson AFB, Ohio. Total costs of the PRIME flight tests, including the costs of the *Atlas* boosters, was $70.5 million.

Another proposed program during the mid-1960s was the short-lived *Aerospace Plane* (not to be confused with the current X-30 *National Aerospace Plane*). Envisioned as a large winged vehicle that would take-off and land horizontally from conventional runways, *Aerospace Plane* would have used a complex multi-phase propulsion system that would, at one point in flight, have extracted air from the upper atmosphere, liquified it, and combined it with stored fuels to become an air-breathing hybrid rocket. *Aerospace Plane* found little support from the AF Scientific Advisory Board, and officially was dropped during 1963.

THE SUCCESSFUL X-15

While the future of *Dyna-Soar* and *Aerospace Plane* were in doubt, one hypersonic research project was reaping invaluable data; the X-15 effort begun a decade earlier. During June 1952, the NACA Committee on Aerodynamics had endorsed a proposal by Bell Aircraft for the development of a hypersonic research aircraft that could study the problems of flight at high altitude and high speed, and solicited concepts from the various NACA Centers and contractors. During October 1953 the AF Scientific Advisory Board recommended development of a hypersonic research vehicle capable of Mach 7 performance. Around the same time, the Navy's Office of Naval Research awarded a contract to Douglas for conceptual studies of a Mach 7+/1,000,000 ft. version of the D-558-2. These various requirements were eventually consolidated into what became the North American X-15.

The X-15 program moved into high gear during 1955 with a series of detailed studies by the NACA, AF and Navy. One baseline effort, by a NACA/Langley team led by John V. Becker, resulted in a configuration bearing a pronounced resemblance to the ultimate X-15 vehicle: an Inconel-alloy aircraft having a cruciform tail with a wedge profile and a low aspect ratio wing. During January 1955 the AF, which had assumed responsibility for administering the design and construction of the X-15 program, briefed potential contractors on the desired characteristics of the aircraft. Although numerous companies initially expressed interest in the program, only Bell, Republic, Douglas and North American actually submitted proposals. Of the four, North American's design appeared most likely to succeed, largely because it had the most cautious and conservative approach. At the end of September 1955, North American received notification that it had won the X-15 contract.

The first of three X-15s arrived at Edwards AFB during the summer of 1959. The X-15 was a striking aircraft, with a polished black Inconel-X superalloy structure spanning 22 ft. 4 in. and stretching 50 ft. 9 in. It was powered by a 57,000 lb. thrust XLR-99 throttleable rocket engine that burned liquid oxygen and anhydrous ammonia propellants. This gave the aircraft 90 seconds of powered flight while enabling it to reach speeds above Mach 6 at altitudes over 60 miles (initial flights were conducted with the much used XLR-11, since the new XLR-99 was not ready).

By the end of 1961, the X-15 already had attained its Mach 6 design goal and reached altitudes in excess of

North American Rockwell studied many designs during the Phase "B" competition. The booster's large air-breathing engines are noteworthy.

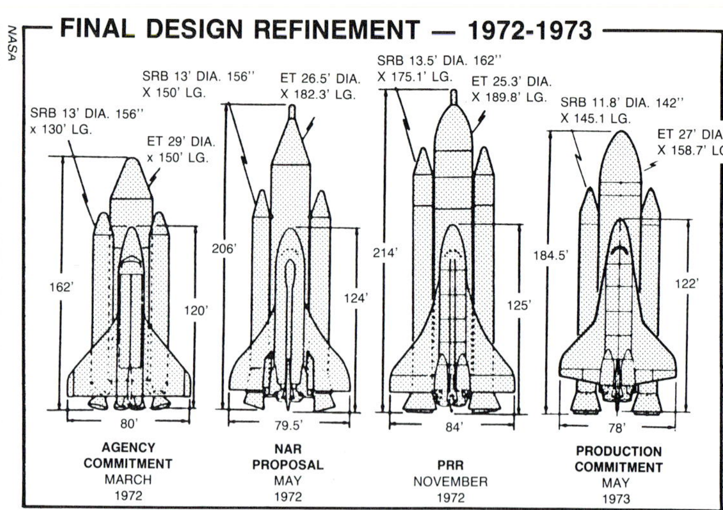

Phase B and ASSC proposals. L to r: Grumman/Boeing study on Saturn S-1C; McDonnell Douglas/Martin Marietta; Grumman/Boeing; North American; and a Saturn V.

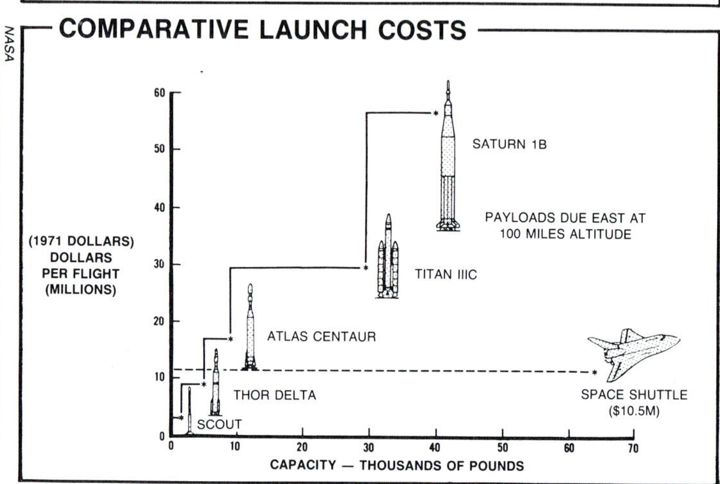

200,000 ft. During August 1962 an inertial navigation system originally designed for the *Dyna-Soar* was installed on X-15 #1 and on August 22, 1963, NASA pilot Joseph Walker took X-15A #3 to an altitude record for piloted aircraft of 354,200 ft. (67+ miles). X-15 pilots repeatedly demonstrated the ability to make precision unpowered reentries and landings. On October 3, 1967 the X-15A-2 set an absolute speed record of 4,520 mph (Mach 6.7) with Major William J. Knight at the controls (this record would stand until the return of the space shuttle *Columbia* from its first mission during 1981). Six weeks later, pilot Michael Adams was killed when the #3 X-15 was destroyed during a high altitude flight. This proved the only fatality during what arguably is the most productive flight research program ever undertaken. Nine more flights were completed (for a total of 199) before the X-15 program was terminated at the end of October 1968.

The X-15 constituted a major step on the path to a lifting reentry spacecraft, examining many of the problems associated with returning from orbit, including heating, aerodynamics, structural loads, physiological requirements, guidance and control. During December 1968, John Becker and the X-15 team were awarded the Eugene Sänger Medal, created to honor individuals and groups who have made significant contributions to the field of reusable spacecraft.

As with many research projects, the X-15 spawned several plans to utilize the craft for purposes other than those for which it was designed. By the mid-1960s, fully 65% of X-15 flights involved using the craft as a carrier for special experiment packages. Several proposals also were forthcoming for possible adaptations of the X-15 vehicle itself. One of these involved rewinging the #3 X-15 with a slender *Dyna-Soar* type wing. This was scrubbed following the loss of the aircraft during 1967. A more radical proposal involved an attempt to build a two-seat X-15B, which would have been launched into orbit on the back of two modified *Navajo* missiles, although thought was given to using a modified *Titan* booster. The X-15B proposals were circulating within North American well before the *Sputnik* launch of 1957, and though the Russian launch stimulated activity on it, no hardware was built, and the proposal never made it further than paper and wind tunnel studies (for the complete X-15 history, see *Datagraph #2* by Aerofax, Inc.).

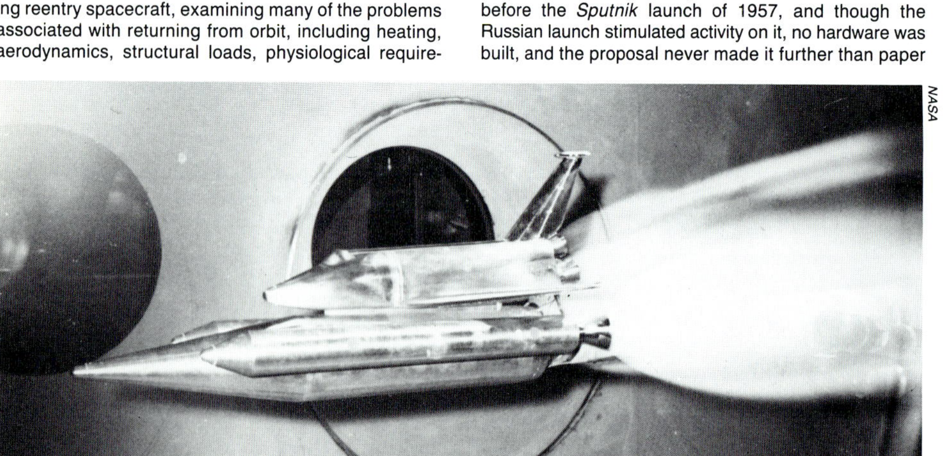

Shuttle exhaust plume studies were conducted using a wind tunnel at NASA's Ames Research Center at Mountain View, California. A 1.9% scale model is shown in the Ames 9 ft. by 7 ft. tunnel at a simulated 69,000 ft. altitude.

OF BLUNT SHAPES AND LIFTING BODIES

During 1951 NACA engineer H. Julian Allen had postulated the blunt-body reentry shape. At the time, the popular conception of a spacecraft was along the lines of the German V2—a slender cylinder with a pointed nose and sweptback fins. Experience showed that any such shape would develop a strong "attached" shockwave streaming from the nose as it reentered the atmosphere, with the associated high heating eventually melting the structure. Allen demonstrated that a blunt shaped object would develop a "detached" shockwave that would carry off most of the heat load, and that residual heating would be well within the capacity of existing materials technology to deal with. Allen's work, first put to use on nuclear warheads of the *Atlas* and *Titan* missiles, formed the basis for the blunt shaped capsules of Project *Mercury*.

With this discovery, one of the major stumbling blocks to developing reentry shapes disappeared. But Allen's symmetrical blunt bodies did have one significant drawback: they followed generally ballistic flight paths that did not permit much variation in landing site. Allen's colleagues at the Ames Research Center realized that the blunt body shape could be modified to generate a small L/D ratio, thus enabling the landing site to be several hundred miles cross-range from the reentry point in the upper atmosphere. This was particularly important to polar operations, where the natural rotation of the earth moves the launch site considerably cross-range during even a short orbital flight.

In short order, three different lifting-body shapes were proposed: the NASA/Ames M2; the NASA/Langley HL-10; and the Aerospace Corporation/Martin A-3. Each of these represented a distinctive design approach. The M2 was basically a modified half-cone, flat on the top with a "boat-tailed" aft fuselage to improve its sub- and transonic aerodynamic performance. It had a round nose and twin vertical fins that had led to its being nicknamed the M2 "Cadillac" as it underwent refinement by an Ames team led by Alfred Eggers. Langley's HL-10 (the "HL" stood for "Horizontal Lander"), developed by a team headed by Eugene S. Love, was a more traditional shape with a flattened and rounded delta wing with sharply upswept tips and a single central vertical tail. The A-3, designed by the Aerospace Corporation and championed by the Martin Company, likewise was a delta with pronounced rounding and twin angled vertical fins. From hypersonic wind tunnel and hypervelocity test facility experiments, researchers predicted each shape had acceptable hypersonic characteristics. However, serious questions remained about their sub- and transonic performance, and whether a pilot could successfully control them prior to the advent of sophisticated computer-based flight control systems. In an attempt to answer these questions, actual prototypes would need to be built and flight tested.

MANNED LIFTING-BODIES

During 1963, engineers at NASA's Flight Research Center (FRC) had built a single pilot plywood M2 shape for a little less than $30,000. This 20 ft. long glider had a fixed undercarriage and was designed to be towed by a Douglas C-47 transport. The first piloted lifting-body flight occurred on August 16, 1963, when NASA research pilot Milton O. Thompson cast off from the C-47 and dropped 10,000 ft. to land on Rogers Dry Lake. This craft, the M2-F1, experienced some handling problems, but offered engineers the expectation that the M2 shape could be flown and landed safely. Encouraged by the results of this limited flight test series, NASA pursued the building of rocket powered lifting-bodies to evaluate their performance up to Mach 1.5 and 60,000 ft. Two competing designs, the M2-F2 and the HL-10, using XLR-11 rocket engines designed for the X-1 a quarter century earlier, eventually reached speeds of Mach 1.8 and altitudes over 90,000 ft. Both were built by the NorAir Division of Northrop Corporation for the budget price of $1.2 million each.

The M2-F2 rolled out of Northrop's Hawthorne, California, plant on June 15, 1965, and was trucked to Edwards the next day. Of conventional aluminum construction, it was 22 ft. long, spanned 9 ft. 7 in., and weighed 4,630 lbs. A retractable tricycle undercarriage was fitted, unusual only in that it was lowered by high-pressure nitrogen gas. A full span ventral flap controlled pitch, while split dorsal flaps controlled roll (lateral) through differential operation and pitch trim through symmetrical operation. Twin ventral fins provided directional (yaw) control and also acted as speed brakes. On March 23, 1966 the M2-F2 completed its first captive flight carried by a NB-52 mothership. The first gliding free-flight, with Milt Thompson at the controls, occurred on July 12, 1966. During the next four months 13 additional glide flights were flown. On May 10, 1967 pilot Bruce Peterson started to exit the second of two planned S-turns only to find the vehicle rolling and banking wildly. Peterson attempted to flare onto the lakebed, but bounced back into the air, then tumbled end-over-end, almost destroying the M2-F2. Peterson was seriously injured (he later recovered and became the director of safety at the FRC.) This was the end of the M2-F2's flight test program, although the airframe subsequently was rebuilt as the M2-F3 incorporating a large central vertical fin in addition to the upswept wingtip fins. Bill Dana took the M2-F3 on its first glide flight on June 2, 1970 and found the handling qualities of this version much improved over its predecessor. On December 13, 1972 the M2-F3 obtained Mach 1.6, its fastest flight, and a week later, on December 21 it reached 71,493 ft., its highest. During the last part of the test program, the M2-F3 was used to check out thruster roll control systems, and analog flight controls—systems that might have potential use in the space shuttle then being designed. The craft was retired after completing 43 flights—16 as the M2-F2 and 27 as the M2-F3; it subsequently joined the collection of the National Air and Space Museum.

Concurrently with the M2 flight test series, the FRC began flying the Langley designed HL-10 shape. This aluminum vehicle was 22 ft. long and spanned 15 ft. across its sharply swept delta wings. The control system consisted of upper body surface and outer fin flaps for transonic and supersonic trim, blunt trailing edge elevons, and a split rudder on the central vertical fin. Bruce Peterson took the craft on its first glide flight on December 22, 1966. During the three minute descent, Peterson discovered he had minimal lateral control over the vehicle, but nevertheless succeeded in landing safely. The NASA immediately grounded it for further study. After wind tunnel tests at Langley, NASA modified the leading edge of the outer vertical fins to direct more airflow over the control surfaces. During its next flight, on March 15, 1968, Jerauld Gentry found the modifications worked well—the HL-10 now handled nicely. The vehicle went supersonic for the first time on May 9, 1969; this was the first supersonic flight of any lifting-body, and an important milestone in the lifting-body program. On February 18, 1970, AF test pilot Major Peter C. Hoag reached Mach 1.86, and nine days later Bill Dana reached 90,303 ft. The HL-10 thus became the fastest and highest flying of the lifting-bodies.

Towards the end of the HL-10 flight test program, NASA embarked on a series of powered landing trials attempting to determine whether the added complexity and weight of landing engines could be justified on the upcoming space shuttle orbiter. It was discovered that the benefits of landing engines, namely the ability to make a missed approach, was far outweighed by the complexity of the installation and the increased pilot workload due to shallower descent angles and higher approach speeds. The HL-10 program was terminated after the 37th flight on July 17, 1970, and its basic shape was deemed the best performer of the lifting-bodies, becoming the basis for several early space shuttle concepts.

While NASA was testing the M2 and HL-10 shapes, the AF embarked on its own piloted lifting-body program. Called PILOT (for Piloted Lowspeed Tests), the vehicle was rolled-out at the Martin plant in Baltimore on July 11, 1967. The X-24A/SV-5P spanned 13 ft. with an overall length of 24 ft. and was basically a scaled-up version of the ASSET SV-5D shape. Empty weight was less than 6,000 lbs., increasing to 11,000 lbs. with fuel and a pilot. Jerauld Gentry completed the maiden flight on April 17, 1969, the first of nine unpowered glides. Gentry also flew the first powered flight on March 19, 1970, reaching Mach 0.87. On October 14, 1970, 23 years to the day after Chuck Yeager's first supersonic flight, John Manke piloted the X-24A on its own initial excursion though the sound barrier. Though it flew well, the X-24A did have one problem: it exhibited a pronounced nose-up trim change that prohibited low-angles of attack during powered flight. FRC engineers concluded that the aerodynamic effects of the rocket exhaust plume impinging on the craft caused the nose-up condition, and warned the designers of the space shuttle to beware of similar problems. The last portion of the vehicle's test program was dedicated to simulating space shuttle approaches and landings.

The X-24A completed its last flight on June 4, 1971, and on January 1, 1972 the AF awarded Martin a $1.1 million contract to modify the vehicle into the X-24B. When the new aircraft was rolled out of Martin's plant, it had grown over 14 ft. in length and 10 ft. in span. Weight was up to 13,700 lbs. The sleek new shape had a double-delta planform swept back 78°, and a nose that was tilted up 3°. Like the other lifting-bodies, it was powered by an XLR-11 rocket engine. John Manke completed the X-24B's first glide flight on August 1, 1973 and its first powered flight on November 15. The vehicle eventually flew Mach 1.76 and reached 97,000 ft.

By the summer of 1975 the space shuttle was well into its operations definition phase, and designers still were debating whether to provide air-breathing landing engines. The primary concern was if low L/D reentry shapes could successfully complete unpowered landings on confined hard runways. John Manke and Mike Love were convinced they could, and on August 5, 1975 Manke guided the X-24B to a perfect landing on Edwards' runway 04. Two weeks later, Love duplicated the performance, both landings occurring within a few feet of the anticipated touchdown point. The X-24B would fly eight more times, being used to familiarize future shuttle pilots with the characteristics of low L/D shapes, before being retired on November 26, 1975. The X-24B, along with the HL-10, proved significant in resolving the final questions of how to fly and land a space shuttle.

Several X-24B follow-on research aircraft were proposed to continue hypersonic research in support of space shuttle and other programs. The Flight Research Center proposed the X-24C, an air-launched vehicle with a Mach 8 potential and 40 seconds of sustained Mach 6+ cruise. Langley proposed the Mach 12 Hypersonic Facilities Aircraft (HYFAC), and the less ambitious Mach 8 High-Speed Research Aircraft (HSRA), while the AF proposed two variations of the X-24B, one rocket powered and one air-breathing.

The proposals were consolidated into a National Hypersonic Flight Research Facility (NHFRF—pronounced "nerf") during July 1976. NASA forecast a $200 million program involving construction of two aircraft with 200 flights over a ten year period. The agency and the AF would start funding the program during 1980 with the first flight during 1983. During 1977 the costs of NHFRF continued to escalate, leading to NASA Headquarters cancelling the effort during September 1977. The NASA acting Associate Administrator for Aeronautics and Space Technology, James J. Kramer, stated that ". . . the combination of a tight budget and the inability to identify a pressing near-term need for the flight facility had led to a decision by NASA not to proceed with a flight test vehicle at this time . . .". It was the same rationale that had caused *Dyna-Soar's* demise ten years earlier. By this time, however, the ultimate winged reentry vehicle—space shuttle—was getting ready for flight.

Martin also proposed a low-speed lifting-body trainer based on the SV-5 shape. The SV-5J, to be powered by a small turbojet, was to be used as an astronaut trainer at the AF Test Pilot School. This vehicle was considered significantly underpowered, and though two airframes were completed, the type never flew. The X-24B, and one of the stillborn SV-5Js (rebuilt to resemble the X-24A) now are on exhibit in the AF Museum.

During August 1972, Milt Thompson and Joe Wells of the Flight Research Center advocated construction of a 36 ft. long, manned "mini-shuttle" to study the proposed space shuttle's Mach 5 to landing performance. Such an aircraft, air-launched from an NB-52, could fly in direct support of space shuttle development, validating wind-tunnel predictions and performance at hyper- to subsonic velocities. FRC estimated that an in-house designed, XLR-99 powered, mini-shuttle could be built for $19.7 million, but critics, ignoring FRC's record of low-cost projects, argued this figure was closer to $150 million. Despite strong backing from Northrop, Martin, and Rockwell, the subscale shuttle succumbed to the cost argument. The actual shuttle's hypersonic, supersonic and transonic performance remained unproven until the first Mach 25 manned reentry from space. Interestingly, no one seems to have proposed a subscale, unmanned shuttle reentry vehicle to be flown like the earlier ASSET and PRIME shapes. In any case, *Columbia* would prove that the aerodynamic performance of shuttle was highly satisfactory.

POLITICAL CONSIDERATIONS

The concept of a reusable space transportation system to make access to low-earth orbit more routine and less expensive started gaining momentum in the U.S. during 1969 as part of NASA's ambitious post-*Apollo* plans. The Space Task Group's report *The Post-Apollo Space Program: Directions for the Future* offered a choice of three long-range plans:

- A $8 - $10 billion per year program involving a manned Mars expedition, a space station in lunar orbit and a 50-person earth-orbiting space station serviced by a reusable ferry, or space shuttle.
- An intermediate program, costing less than $8 billion per year, that deleted the lunar-orbiting space station, but kept the other elements.
- A relatively modest $5 billion per year program that would embrace an earth orbiting space station, and the space shuttle as its link to earth.

When the White House rejected these proposals, NASA was left with identifying a project that could both gain enough political support for approval, and still be sizable enough to keep its development engineers usefully occupied during the 1970s. The space shuttle emerged as that project because it was the logical first step to implementing any future plans for a Mars mission or space station.

Once it had decided that the shuttle was its top priority program, NASA during the fall of 1970 unsuccessfully attempted to get it approved by the White House. A new administrator, James Fletcher, was nominated to head the agency during the spring of 1971, and was determined to get the shuttle initiative approved that year. NASA calculated that in order to get White House and congressional approval to develop the new space transportation system they would have to significantly reduce the cost of access to space, and get the DoD to commit to use the system for all of its launch needs once it became available. Thus, providing a shuttle that met DoD's requirements became a key element in NASA's strategy for program approval. In addition, developing easy access to space was the first step in NASA's ultimate goal, the development of a permanent manned space station.

While meeting DoD requirements implied a large, high-performance shuttle, NASA also had to meet its commitment to the Office of Management and Budget (OMB) to make access to space more economical. The shuttle was the first NASA manned space program to be subjected to formal economic analysis and requirements. After an

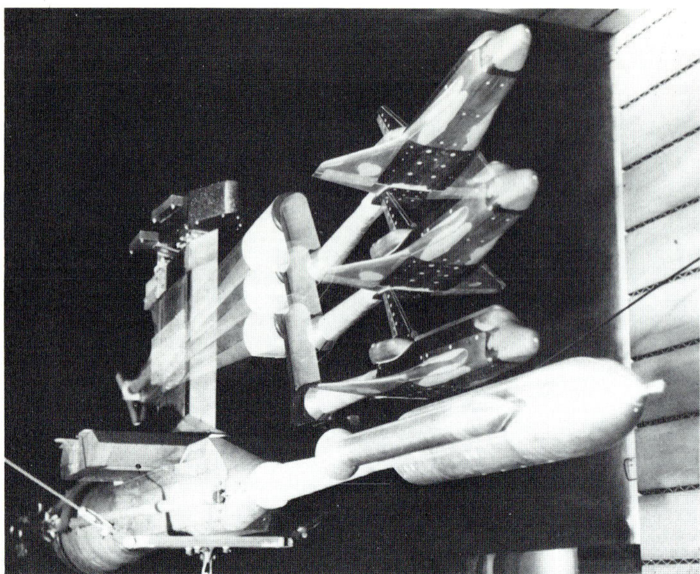

NASA

Ames Research Center 14 ft. transonic tunnel was utilized during 1974 to conduct simulated Shuttle mission aborts at transonic and supersonic speeds.

Rockwell International

A .36 scale wind tunnel model is seen during pre-delivery preparations at Rockwell's Downey, California facility. Model length was 38.71 ft.

internal NASA study failed to impress the OMB, contracts were issued during June 1970 to the Aerospace Corporation for estimating payload and launch vehicle costs; to Lockheed for analyzing the impacts of shuttle capabilities on reducing payload delivery costs, and to Mathematica for a comparison of total costs of carrying out likely future NASA, DoD and commercial space missions using the shuttle and other launch vehicles. NASA Deputy Administrator George Low later commented that Mathematica's analysis impact on the shuttle decision was ". . . influential and unfortunate . . .". This was because although the analysis concluded the shuttle would indeed significantly reduce the costs of launching payloads, the existence of a public study making such claims forced NASA to publicly maintain that the shuttle was a good investment on economic grounds.

It became clear even during NASA's early in-house analyses that any economic justification depended crucially on the shuttle being the only U.S. launch vehicle during the 1980s. In particular, NASA had to gain agreement from the national security community to use the shuttle to launch all military and intelligence payloads, which were projected to be roughly a third of all future space traffic. Thus DoD support of the shuttle was crucial on both political and economic grounds.

Accommodating all DoD missions required a shuttle that could handle payloads up to 60 ft. long, could launch 40,000 lbs. into polar orbit, and 65,000 lbs. into a due-east orbit. Of the Shuttle's critical design parameters, only the maximum payload width of 15 ft. was based primarily on a NASA requirement; a projection that any future space station would be built using modules of that diameter. But the requirement that had the most impact on the shuttle design was that for high cross-range—the ability to maneuver upon reentry to either side of the vehicle's ground track. The AF wanted a 1,100 to 1,500 mile cross-range capability to allow a quick return from orbit to secure military airfields. In particular, the AF wanted to be able to launch the shuttle into polar orbit from Vandenberg AFB, California, have it rendezvous with an already orbiting reconnaissance satellite, service it, and return after a single revolution to Vandenberg. The landing strip would have moved some 1,100 miles to the east as the earth rotated during the 90 minute flight.

Even though DoD requirements drove important aspects of the shuttle design, it was not clear how strong military interest was. Robert Seamans, Secretary of the AF, saw ". . . no pressing need . . ." for the shuttle, but characterized it as ". . . a capability the AF would like to have . . ." Thus although it drove several significant shuttle design requirements, the AF was not totally committed to the concept of a space transportation system and would continue to develop and purchase its own expendable (*Titan* and *Atlas*) boosters. The AF also did not want to contribute any substantial amount of money to shuttle development, other than funds to build a shuttle launch complex at Vandenberg AFB. During 1971 the AF did agree not to develop any new expendable boosters, and leaders of the national security establishment communicated their support of the shuttle program both to the White House and to Congress. This decision, to replace all launch systems with the shuttle, would have significant impact to the DoD, and everyone else desiring access to space, after the 51-L accident during 1986.

NASA

A 1/10th scale "MSC 12.5K Space Orbiter Shuttlecraft" while suspended from a Sikorsky CH-54A prior to a successful drop test at Ft. Hood, Texas on May 4, 1970. Tests verified high AoA recovery to normal flight attitude.

CHOOSING A CONFIGURATION

During 1969, NASA awarded "Phase A" contracts for the study of an "Integral Launch and Reentry Vehicle" (ILRV) to General Dynamics/Convair, North American Rockwell, Martin-Marietta, McDonnell Douglas, and Lockheed Missiles and Space Company. But even within NASA there was no firm consensus of what a space shuttle should look like. At the Manned Spacecraft Center (MSC—later to be JSC), engineers led by *Mercury* capsule designer Max Faget wanted a huge, fully reusable winged vehicle that strongly resembled Krafft Ehricke's orbital version of *Bomi*. In fact, the only significant difference was that Faget's orbiter drew heavily on the lessons learned from the HL-10 program. Other NASA centers supported partially or completely recoverable versions of the *Saturn* family of boosters, and there was still some minor support of a single stage to orbit (*Aerospace Plane*) vehicle. The contractors had equally as many ideas.

GENERAL DYNAMICS/CONVAIR

Convair released their final technical report on October 31, 1969. Their studies centered around two different vehicles, designated FR-3 and FR-4, that NASA had indicated an interest in during the mid-term review held earlier in the year. Both vehicles had a 15 ft. by 60 ft. payload bay and were capable of lifting a 50,000 lb. load to the required 270 NM orbit. Late in the study, NASA asked about the feasibility of carrying a 22 ft. by 30 ft. payload, and Convair made additional studies that concluded the FR-4 could be modified relatively easily for the task, while the FR-3 would require considerable work.

The shape of the FR-3 orbiter, and of all the FR-4 components, was adopted directly from the FR-1/T-18 vehicle GD/Convair developed early during the ILRV studies. The FR-1/T-18 vehicle was a system of three identically-sized elements ('triamese') with all engines burning at lift-off and propellant cross-fed between the elements. The body consisted of a constant section with a tapered forebody terminating in a hemispherical nose. The forebody shape was parabolic in planform with a partially straight-lined lower surface ramp in the profile view. The constant section consisted of a flat-bottom with sides that tapered inward at a 12° angle and a full upper radius that was to allow maximum usage of state-of-the-art cylindrical or truncated conical cryogenic dewars. The flat-bottom improved the hypersonic lift-to-drag ratio, and and also allowed easy storage of the switchblade type subsonic wings. The sides sloped inward toward the top to improve the hypersonic L/D, and to reduce entry heating on the side surfaces. The vee tail was attached high on the afterbody to provide hypersonic and transonic stability. Hypersonic roll control was to be achieved via differential deflection of the ruddervators on the vee tail.

The general arrangement of both (FR-3 and FR-4) orbiters consisted of a nose compartment 14 ft. long that had accommodations for a two-man crew in conventional side-by-side seating, and contained all avionics. Immediately aft of the pressure bulkhead which terminated the crew compartment was a 7 ft. long subsystems compartment housing environmental, electrical and hydraulic subsystems. The air-breathing engines were housed immediately aft of the subsystems compartment, but forward of the LO_2 tank. It was projected two Rolls-Royce RB211-22 turbofan engines would be rotated to flying position by a hydraulically-actuated double-acting mechanism similar in design to the wing-fold mechanism of carrier-based fighter aircraft. The main LO_2 tank formed an integral part of the airframe and was to use internal frames and external stringers with monocoque domes on the ends. The FR-4 was provided with 10,075 cubic ft. of LO_2 storage, while the smaller FR-3 had 7,618 cubic ft. The area below the LO_2 tanks was used for small on-orbit propellant tanks. Behind the LO_2 tank was the 15 ft. by 60 ft. payload bay. The wing-pivot bulkhead was located approximately mid-bay, and supported the variable-geometry wings with large clevis fittings at the outboard end of the carry-through truss structure. At the forward and aft ends of the payload bay, heavy bulkheads were to support the two nose and two aft landing gear assemblies. These installations were to be completely outside the basically circular structure of the payload compartment. The nose and aft landing gear concept was very similar to the B-52 arrangement, except the wide track (26 ft.) negated the need for outriggers. The landing gear could not be extended until after the wings were deployed, but this was not considered a disadvantage, since no attempt to make a gear-down landing with the wings stowed was anticipated. The wing was deployed using screwjacks driven by hydraulic motors in a fashion similar to the F-111 variable sweep system. The main LH_2 tank was located directly aft of the payload bay, and was similar in construction to the LO_2 tank; 19,140 cubic ft. of LH_2 storage was anticipated for the FR-4, and 15,050 for the FR-3. The compartment aft of the LH_2 tank contained the thrust structure, engine gimbal support, stabilizer attach structure and propellant lines with a non-structure outer fairing surrounding the entire compartment. Three 400,000 lb. thrust (sea-level) high-pressure bell-nozzle rocket engines were supported at gimbal points on the thrust structure. The engines were protected during reentry by a body-flap extension of the lower-fuselage.

NASA

Two Grumman "Gulfstream IIs", N-946NA (shown) and N-947NA, were acquired by the NASA during November 1976 for use as Shuttle Training Aircraft. They are used for crew training, weather monitoring, and approach practice.

Other design details on the orbiter included a 30 in. diameter pressurized tube that provided access for the crew from the flight-deck to the payload bay. This tube was routed from the pressure bulkhead at the rear of the crew module, along the bottom centerline through the subsystem compartment, to the bulkhead at the aft end of the air-breathing engine compartment. At this point, it swung out around the lower forward LO_2 tank dome, along the lower left side of the compartment between the lower structure and the LO_2 tank. It then swung up around the aft LO_2 tank dome and into the payload bay. The payload bay end terminated in a flexible 42 in. diameter tube that could be attached to a payload to provide shirt-sleeve access.

The basic structural materials were aluminum alloy for the propellant tanks and main body structure, titanium alloy for the lower heat shield supports and thrust structure, and boron-aluminum composite in longitudinal stress areas such as beam caps and stiffeners. The projected thermal protection system consisted of micro-quartz and Dynaflex insulation external to the basic structure with post-supported cover panels, primarily of cobalt alloy on the lower surface and 811-titanium on the upper and side surfaces.

The FR-3 booster was considerably larger than either the FR-1/T-18 or FR-4 boosters since the entire boost function was contained in one vehicle instead of two in the other designs. The booster was generally derived from the FR-1 shape, but had a blunter nose and steeper side slopes. The cross section was still flat-bottomed with a full upper radius to accommodate the largest possible cylindrical tank. Sufficient space was reserved in the nose for the four RB211-56 air-breathing engines and their JP-4 fuel, and various subsystems. The two man crew was situated in a compartment elevated above the nose section to improve visibility during landing operations. Again, a vee tail was used, and no elevons were fitted, roll control being achieved with the ruddervators. The LO_2 and LH_2 tanks were constructed of 2021 or 2022 aluminum alloy with a thermal protection system of 811-titanium fairing panels over micro-quartz insulation.

The FR-4 boosters generally were similar to the orbiter, except the payload bay was omitted, and the LO_2 and LH_2 tanks shared a common, center bulkhead to maximize propellant storage. The booster stages also were larger, and housed nine 400,000 lb. th. main boost engines each. Three air-breathing RB211-56 engines were needed on each stage to handle the 356,000 lb. flyback weight. In this system, one booster was mounted on each side of the orbiter, and all three stages were ignited upon lift-off.

Various studies of fixed-wing alternatives were made, with the conclusion that there was little overall weight difference between fixed and stowed wings. Because of the unknown transonic stabiity problems that might be encountered by a fixed-wing configuration, all configurations proposed by GD/Convair used switchblade type stowed-wings. The studies further indicated that delta wings, or

NASA

Contingency landing operations were rehearsed using a mock-up of the shuttle cockpit section moored in the Turning Basin at Complex 39 to simulate shallow water ditching.

NASA

Full-scale shuttle mock-up during April 1974, at Rockwell's Downey, California facility. Noteworthy is the early style forward RCS mounted behind the doors on the nose.

Mock-up (named "Pathfinder" and tagged "098") was modified by Teledyne-Brown in Huntsville, Alabama during 1977 for display in Japan.

The engineering and development mock-up facility is a full-scale, high-fidelity mock-up of the Shuttle housed in Building 9-A of the Johnson Space Center.

double deltas, would reduce the aerodynamic center shift during transonic transition, but they were heavy and inefficient during subsonic cruise back to the launch site since their aspect ratio was low.

NORTH AMERICAN ROCKWELL

The North American Rockwell Phase A studies were divided into two parts: The first two months encompassed comparisons of reusable logistic vehicles with expendable boosters. The second part involved investigations of a two-stage reusable vehicle using similarly shaped booster and orbiter elements with fixed, low-mounted, swept wings. These vehicles could be configured to deliver 10, 25 or 50,000 lb. payloads into a 270 NM orbit inclined at 55°, and were designed for vertical take-off, with the orbiter mounted forward on the booster. This orbiter location provided a forward center of gravity during the ascent phase of the mission. The design approach was to design vehicles with a large planform area, and flat base, that entered the atmosphere at a high angle of attack to minimize structural heating, thus minimizing the requirements for a thermal protection system. The mated system would weigh 4,476,000 lbs. at lift-off.

The orbiter was a swept wing vehicle spanning 146 ft. with 2,830 square ft. of effective wing area. A fuselage 202 ft. long contained LO_2 and LH_2 tanks a 15 by 60 ft. payload bay, room for ten passengers, and a crew of two. Two 510,000 lb. th. (sea-level) boost engines, with two-position nozzles and a maximum expansion ratio of 120:1, were mounted side-by-side in the aft fuselage. The main boost engines, operating at 10% thrust, also were to be used for orbital maneuvers. Four JT3D-7 turbojet engines, each rated at 19,000 lbs. th., were provided for landing and ferry operations.

The payload bay was mounted at the orbiter's center of gravity to provide a vehicle that could be trimmed on reentry and during cruise with, or without, a payload. The location of the payload bay required the use of two LH_2 tanks, one forward of the bay, the other aft. The forward LH_2 tank was integral with the LO_2 tank, the propellants being separated by a common bulkhead.

The thermal protection scheme chosen was considered the lightest possible. Propellant tanks, constructed of fusion welded aluminum, were suspended and therefore not subjected to the vehicle's aerodynamic and thermodynamic loads. The primary load carrying structure was a skin-stiffened titanium shell, which in order to withstand the reentry thermal environment, was protected on the bottom and lower sides of the fuselage by a densified quartz external insulation. Bare titanium was considered satisfactory on the top of the orbiter because of the milder thermal environment. A fiberglass insulation was used inside the orbiter to minimize heat transfer to the propellant tanks.

Shuttle full-scale aft fuselage mock-up during August 4, 1972 at Rockwell's Downey, California facility.

The 280 ft. long booster used a propellant tank constructed of 2219A1 aluminum alloy as the primary load-carrying structure. The LO_2 tank was located forward of the LH_2 tank with a common bulkhead separating the propellants. An aluminum truss and reinforced phenolic/polyimide honeycomb-sandwich substructure was provided to maintain the desired flat base area. This substructure also was used to bond densified quartz external insulation to the booster's bottom and sides. The 244 ft. span wings were constructed of titanium. Booster main propulsion was to be eleven high-pressure bell-nozzle engines similar in type to the orbiter's engines. Four JT9D-15 turbojet engines would burn 57,800 lbs. of JP-4 fuel during a 310 mile cruise back to the launch site.

It was anticipated that design and production of development hardware would consume 4.5 years, after which six flight vehicles would be built at the rate of two per year. A launch rate of 50 per year was anticipated using the six vehicles. Two-hundred direct support personnel would be required for maintenance and launch operations using airline type scheduling.

MARTIN-MARIETTA

Perhaps the most unusual proposal to be generated out of the Phase A studies was the Martin-Marietta *Spacemaster*. Like the rest of the Phase A contractors, Martin proposed a fully reusable system based on a separate orbiter and flyback booster. It was, however, the nature of the booster that was unusual.

The orbiter was a severely swept blended double-delta planform spanning 107 ft. The tips of the wings were canted almost straight up to act as vertical fins. The 181.2 ft. long fuselage housed the large boost LH_2 tank forward of the payload bay, which could accommodate payload canisters measuring either 22 by 30 ft. or 15 by 60 ft. Maximum payload weight was 25,000 lbs. Twin LO_2 tanks were in the blended section of the wing/fuselage on either side of the payload bay, on the vehicle's center of gravity. Smaller LH_2 and LO_2 tanks for use on-orbit were carried aft of the main LO_2 tanks. Two 400,000 lb. th. boost engines were housed in the aft fuselage and a crew of two was carried in the extreme nose. A landing speed of under 180 knots was projected, as was a cross-range capability of 1,700 miles. This version of the orbiter had no provisions for air-breathing landing engines.

The baseline first stage booster consisted of twin fuselages, 28 ft. in diameter and 197 ft. long. These were connected by a forward wing that spanned 70 ft. between the fuselages and contained eight gaseous hydrogen (GH2) burning air-breathing engines. A rear wing had a total span of 148 ft. with 20° of anhedral on the outer panels and used a NACA 2415 section. A total of fourteen (seven per fuselage) 400,000 lb. th. LH_2/LO_2 boost engines were provided. A crew of two was carried in the extreme front of the left fuselage. Each fuselage was composed of an integral aluminum LO_2 tank forward of the front wing, and a cylindrical integral LH_2 tank between the front and aft wing. Reaction control systems were carried forward of the LO_2 tank, the front landing gear was between the two tanks below the front wing, and the main engines and rear landing gear were carried behind the LH_2 tank. Twin sweptback vertical tails rose 24 ft. above each fuselage. The orbiter was suspended from the forward and aft wings, and the system had a gross lift-off weight of 3.5 million lbs.

A larger version of the same design, this one with a booster 204 ft. long, spanning 160.8 ft. and powered by 16 boost engines also was proposed. The orbiter for this version would have four liquid hydrogen burning Rolls-Royce RB-162 landing engines under the forward section of the payload bay that could be lowered into the airstream when needed. This orbiter would have been slightly larger, growing to 186.8 ft. long and spanning 110.3 ft., and using three boost engines instead of the baseline two. This version would have had a gross lift-off weight of 4,146,000 lbs. with a payload capacity of 50,000 lbs.

Other booster designs were studied, all based on a single conventional fuselage, with a variety of wing configurations including variable geometry, straight, delta, double-delta and tapered aft wings with forward canards. The alternative booster chosen for study was the tapered aft wing/forward canard configuration with 13 boost engines. This vehicle would have been 258.5 ft. long with a 156 ft. wingspan and a large vertical tail that was located centrally above the aft wing. Several positioning and mating surface options for the orbiter/booster also were studied with booster-top-to-orbiter-bottom ultimately being selected. This configuration, although the second choice during Phase A, would become the favored approach for the McDonnell Douglas/Martin Marietta team during Phase B.

McDONNELL DOUGLAS

The McDonnell Douglas Phase A study was completed during November 1969 emphasizing a "point" two-stage-to-orbit vehicle, with several other configurations examined based on parametric excursions from the "point" design. The overall design goal was to minimize operational and program recurring costs and was to be achieved through high system reliability, vehicle recoverability and rapid ground turnaround made possible through modular component design and use of an integrated onboard self-test and checkout system. A 25,000 lb. payload could be delivered to, and returned from, orbit in a 15 by 30 ft. payload bay. The total system had a gross lift-off weight of 3.4 million lbs.

Thirteen configurations, all with an orbiter based on the HL-10 lifting-body shape, were considered by McDonnell Douglas during Phase A. The majority of these were intended to find the ultimate booster configuration, and the best mounting location for the orbiter (on top, on bottom, semi-submerged, etc.). Of the thirteen, only five were deemed worthy of further study: a large lifting-body booster with the orbiter mounted semi-submerged in the bottom; a variable geometry wing booster with the orbiter mounted recessed in the top; a large delta-winged booster with the orbiter mounted top-to-bottom; and another variable geometry wing booster with the orbiter mounted top-to-bottom; and a large clipped-delta booster with the orbiter mounted bottom-to-bottom. This last configuration was the one chosen for detailed study.

The orbiter was a 107 ft. long vehicle based on the HL-10 shape. The aft-fuselage base area was modified slightly to accommodate two 415,000 lb. th. boost engines. The launch propellant tanks were integral with the primary body structure to maximize volume. The geometric center of the payload bay was located on the vehicle center of gravity to permit the option of returning from orbit with or without the full payload. The boost LO_2 tank was forward of the payload bay with the tank skins conforming to the mold shape on the sides and bottom. Forward and aft tank bulkheads provided vehicle structure. The LH_2 tankage was composed of three tanks, the two side tanks providing mold line skin on the sides, top and bottom. The inboard walls of these tanks also provided a thrust structure for the boost engines. The third LH_2 tank was located aft of the payload bay, and acted as a collector for the boost engines. A two man crew would be housed in a shirt-sleeve environment in the extreme nose of the orbiter. The orbiter thermal protection system consisted of titanium, Renė-41, nickel-chrome (NiCr) and columbium (Cb-752) shingles (tiles). An alternate configuration was considered using hardened compacted fibers (HCF) in place of the nickel-chrome and columbium shingles. The HCF shingles were not recommended since the ones evaluated were heavier than the equivalent alloy shingles and not as durable. Columbium shingles were preferred over nickel-chrome because of the extensive experience acquired in their manufacture and use during the ASSET program. Alternate orbiter configurations, still based on the HL-10, were developed that allowed 50,000 lb. payloads of either 15x60 ft. or 22x30 ft. dimensions.

The carrier was a 195 ft. long clipped-delta configuration with ten boost engines identical in type to those on the orbiter. Dual cylindrical propellant tanks formed the primary body structure. The 15% thick delta wing contained the landing gear and air-breathing engines and their JP-4 fuel. The LO_2 tank was in the forward fuselage to minimize engine gimbal angles for initial thrust vector and center of gravity tracking during boost, and also to locate the empty vehicle c.g. further forward. The aft bulkhead was convex for minimizing residual propellants. The vehicle body structure was provided by the propellant tank walls with integrally machined rings and stringers. A two man crew was housed in the forward fuselage ahead of the LO_2 tank. The booster would use structural titanium panels with Renė-41 leading edges for the wings and tail. The rest of the booster would use titanium and Renė-41 shingles for thermal protection.

Operational mission profiles called for the mated vehicles to be launched vertically powered only by the booster's engines. The vehicles would separate at an altitude of approximately 35 n. miles at a velocity of 9,000 ft. per second. The orbiter's boost engines then would be ignited for insertion into a 45 by 100 nm orbit, which would then be circularized. After proper phasing, the orbiter would be raised to the 270 nm mission altitude. Reentry would be accomplished at 25,000 fps at a flight path angle of −1.5°. A banked attitude and 50° angle of attack would be maintained during pull-out until the lower surface temperature reached 2,200° F and this temperature would be maintained by bank angle modification until normal load factors became restrictive. A roll maneuver to wings level would take place at speeds between 6,000 and 10,000 fps in order to stretch the range and satisfy the 390 nm cross-range requirement. When the velocity reached 3,900 fps, the orbiter was pitched down and allowed to descend using the profiles developed by the HL-10 during landing trials at Edwards. After separation, the booster would pitch to 50° angle of attack while inverted and continue in this mode through ballistic apogee. A roll-out to upright flight then would be performed to allow pullout above the thermodynamic boundary. Upon reaching subsonic horizontal flight, the booster would deploy the air-breathing engines and be flown back to the landing site.

Dennis Jenkins

"Pathfinder" was placed on permanent display at the Alabama Space and Rocket Center, Huntsville, Alabama, during 1988. Never utilized black SRBs have filament wound cases designed for special missions from Vandenberg AFB.

It was estimated that the development program, to include building two flight-test orbiters and two boosters, would cost $5,946,000,000 and take 21 months. At a launch rate of 12 flights per year, using three orbiters and two boosters, the cost-per-pound to orbit would be $119. If this was increased to 100 flights per year, the cost would drop to $67 per pound.

Studies also were conducted on finding an all-azimuth launch site to replace both the Kennedy Space Center and Vandenberg AFB. Over 100 sites were examined, with McConnell AFB, located just outside Wichita, Kansas, being selected. Its location was such that no serious geographic constraints would be encountered (mountains, lakes, etc.) and adequate abort landing sites were located along probable launch routes. The McDonnell final report recommended that further studies be performed, particularly of climatic conditions, to assess the advisability of an entirely new launch facility.

LOCKHEED

The Space Systems Division of Lockheed Missiles and Space Company (LMSC) released their Phase 'A' report on December 22, 1969. The study reflected LMSC's 10 years of interest in reusable launch and space operations concepts, during which the company had expended significant resources studying lifting-entry spacecraft configurations and heat transfer technologies. Following trade study evaluations which led to the final requirements and design parameters, four baseline vehicles were sized on the basis of two-stage and dissimilar triamese concepts, with payloads of 25,000 and 50,000 lbs. each.

Although the LMSC study, led by Wilson B. Schramm, contained an in-depth analysis of two (two-stage and triamese) 25,000 lb. payload concepts, it concluded ". . . the best insurance against development risk would seem to be an approach based on large system [50,000 lb.] size with growth potential. . .". With this in mind, the report centered on a 50,000 lb. payload orbiter. The orbiter could accommodate payloads measuring either 15 x 60 ft. or 22 x 30 ft., and offered a potential for a 1,500 n. mile cross-range, but Lockheed noted that by reducing this to 400 n. miles, between 8,000 and 10,000 lbs. could be shaved off the thermal protection system. The two-stage configuration was preferred over the dissimilar triamese since it displayed significant advantages in reliability, safety and also would cost 20% less to operate.

Lockheed made extensive use of a system synthesis computer model called "MAGIC" that allowed the interrelated effects of various parameters to determine launch system sizing. As an example, orbiter thermal protection system weight is strongly influenced by wing loading, which in turn is derived from lifting body planform geometry in relation to weight of the primary structure and subsystems involved in performing both the ascent and on-orbit mission functions. MAGIC allowed multiple "what-if" scenarios to be considered in achieving optimal system sizing. The program had been in use at LMSC for about two years, and had yielded generally good correlation between models, wind tunnel testing and flight tests.

The 50,000 lb. two-stage booster design was 237 ft. long, weighing 322,500 lbs. empty. It featured low mounted delta wings that swept up at the tips to form vertical stabilizers. Liquid hydrogen was carried in a cylindrical tank midships, with liquid oxygen in a conical tank forward. A crew of two was to be carried in the extreme nose. Thirteen 400,000 lb. th. engines were mounted high in the fuselage to minimize center-of-gravity offset effects with the orbiter attached. Four

Dennis Jenkins

Ex-American Airlines Boeing 747-123, N-905NA shortly after launching the first ALT free-flight. Chase aircraft is Northrop T-38A, N-961NA.

Erik Simonsen

Boeing 747-123, N-905NA, carrying OV-099 "Challenger" during a return trip from Edwards AFB, California to Kennedy Space Center, Florida.

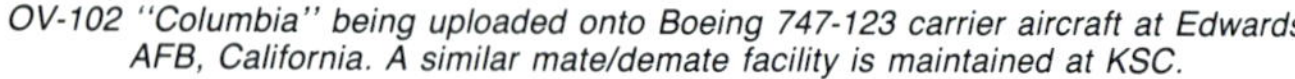

Dennis Jenkins collection

OV-102 "Columbia" being uploaded onto Boeing 747-123 carrier aircraft at Edwards AFB, California. A similar mate/demate facility is maintained at KSC.

NASA

"Pathfinder's" original configuration prior to more defined shuttle mock-up. As seen, dimensions, weights, and balances were similar to the actual vehicle.

38,600 lb. th. turbofan air-breathing engines were on fixed mounts on the extreme rear fuselage. JP-4 fuel for the air-breathers was carried in four tanks contained in the wings.

The delta lifting body orbiter featured non-integral cryogenic tankage which was thought to have less manufacturing risk, and offer better inspection and maintenance characteristics. A graceful blended delta again ended in wingtips up-swept to form vertical tails. The relatively large fins improved the subsonic lift-to-drag ratio, as well as hypersonic stability. The extreme nose contained a crew module with seating for four, followed by an LH_2 tank containing 46,508 lbs. of fuel. Two kidney shaped LO_2 tanks, each containing 145,404 lbs. of oxidizer, were below, behind and outboard of the LH_2 tank, flanking the payload bay. Additional LH_2 was carried in two tanks behind the payload bay.

Both the orbiter and the booster were to be of conventional aluminum construction. A Tantalum-Tungsten alloy (Ta-10W), capable of withstanding 2,750° F was to be used for the orbiter nose cap. Heat shields of LI-1500, a Lockheed developed silica based insulation material, were to be bonded directly to the orbiter and booster structures. (Interestingly, this material later would evolve into the "tiles" used on the production space shuttles). Alternately, a combination of TD-NiCr/Cb-752 and René-41 corrugated heat-shields could be post-supported with Dynaflex or micro-quartz insulation packaged between the shield and the structural panel.

By using a large number (13) of engines in the booster, it was felt that the probability of total mission success was improved. Simulations showed that two engines could be lost as early as 10 seconds after lift-off and the system still could attain orbit if the remaining 11 engines were throttled to 115%. An abort-to-orbit could be achieved under the same conditions with as many as five engines out. The LMSC study noted that the probability of total mission success was 0.995, with catastrophic engine failure being the predominate concern. The analysis concluded that there would be 190 catastrophic failures (i.e.; loss of crew and vehicle) per one million flights.

A total of five orbiters and five boosters would be required for the flight test program with each stage (orbiter and booster) flying 175 horizontal test flights, and 25 vertical flights. Two of the orbiters and two boosters would be refurbished for later use in the operational fleet. Total RDT&E costs were estimated at $5.5 billion. It was anticipated that vehicle turn-around normally could be accomplished in sixteen 8-hr. shifts, with the possibility of expediting it to 10 shifts if required. Based on 1,000 flights, the total recurring cost per flight was predicted to be $1,255,000, with the operating cost per pound of payload to be $15.10 (an additional $10 per pound was required to amortize RDT&E costs). To support this flight schedule, Lockheed proposed building five orbiters at $61 million each, and two boosters at $98 million each, to supplement the vehicles refurbished from the development testing.

One ironic feature of the LMSC study was to point out a discrepancy between the NASA operational cost model and the ILRV cost model. Using the STS model, a 1,500 man work force was predicted for the operational era, instead of the 400 man force predicted by the ILRV model. Lockheed noted it was ". . . hard to envision such an expenditure . . ." for an operational system. (Today, Lockheed Space Operations Company employs over 6,000 people to process the shuttle system at KSC.)

The Lockheed study believed the basic aerodynamic technology existed to build a shuttle, but recommended additional study on orbiter landing modes and aborts; providing a larger cargo bay than either the 15 x 60 ft. or 22 x 30 ft.; systems redundancy; and the use of LH2 as a fuel for the air-breathers instead of JP-4.

DISSENSION WITHIN NASA—THE MSC DC-3

On January 23, 1970 the Manned Spacecraft Center (MSC) in Houston initiated an in-house design study of space shuttle concepts. The study was initiated due to growing concerns over the development of requirements for the upcoming Phase B Request for Proposals (RFP). The specific concerns centered around the lack of convergence of the design requirements, the high development and total program costs and risks, and the long development time. The basic philosophy of the MSC study was to lower the development cost and risk, and to develop a system that could be operational by 1975—the year the last *Apollo* flight was scheduled. The study was variously known as the MSC-12.5k/MSC-15k (denoting payload capacity), and ultimately as the MSC DC-3 because of its relative simplicity.

To accomplish these goals, the engineers at MSC designed an orbiter with an 8 by 30 ft. payload bay capable of carrying 15,000 lbs. Both the orbiter and fly-back booster would make use of as much existing spacecraft and aircraft hardware as possible to minimize new development time and costs. The orbiter would have a low cross-range (200 mile) capability, and the booster initially would be recovered down-range, then flown back to the launch site.

The vehicle emerged in an April 27, 1970 report as two vertically launched, fully reusable stages with the orbiter mounted to the top of the booster in a long overlap, or "piggy-back", arrangement. The orbiter contained two boost engines, two on-orbit maneuvering engines, all burning LO_2/LH_2, and six air-breathing landing engines burning conventional JP-4. Both the LO_2 and LH_2 tanks were carried internally below the payload bay. JP-4 was carried in an aircraft style wet wing. An aluminum and titanium airframe was protected by silica based thermal protection tiles (very similar to those used today) and carbon-carbon laminates. The general configuration was conventional with a straight wing spanning 90 ft. 10 in. with a 14° leading edge sweep. A conventional tail arrangement was provided with the horizontal stabilizer spanning 43 ft. 8 in. with a 10° leading edge sweep. The vertical tail stood 41 ft. 8 in. high and was angled back at 45°. The six air-breathing engines were housed in titanium pods on top of the wings at roughly mid-span.

The fly-back booster was a straight-winged vehicle spanning 141 ft. with a 14° sweep on the leading edge providing 2,840 square ft. of area. A 203 ft. long fuselage housed 11 main propulsion engines of the same type as used by the orbiter. Four advanced technology turbofans were mounted on top of the wings in nacelles having closures to protect them during boost and reentry. A crew of two was provided.

The reference mission included a circular orbit at 270 n. miles with a 55° inclination. The on-orbit lifetime for the orbiter was 7 days. A launch rate of 30 flights per year was projected with a fleet of six orbiters and four boosters with each orbiter and booster having a life of 100 missions. Both the booster and the orbiter had automatic approach and landing equipment, and the system was to be capable of launch on 48 hours notice for space station rescue missions.

Several one-tenth size dynamically scaled models of the MSC DC-3 (12.5k) orbiter were drop tested from an Army CH-54 helicopter during 1970. The initial tests commenced on May 4, 1970 at Ft. Hood, Texas, and later were continued at the White Sands Missile Range in New Mexico. The tests were designed to demonstrate the orbiter's transition from a high angle of attack reentry to a level cruise attitude; the stability of the vehicle in stalled conditions; and to obtain free-flight data to assist in aerodynamic analytical transition prediction techniques. The model was about 13 ft. long, with an 8 ft. wingspan, and weighed about 600 lbs. Constructed of aluminum and fiberglass, it was dropped from altitudes up to 12,000 ft., and used a parachute recovery system.

PHASE B

After evaluation of the various Phase A proposals, NASA let two contracts on July 6, 1970 for continued study of the Integral Launch and Reentry Vehicle. The companies selected to work on "Phase B" included a team consisting of McDonnell Douglas and Martin-Marietta; and North American Rockwell, who was joined by General Dynamics/Convair as a risk sharing subcontractor. This phase was to involve a more detailed analysis of mission profiles, and the vehicles needed to complete them. Economics was continuing to play an important role in the Shuttle's design, but the primary consideration was beginning to shift from providing the most economical operational system, to reducing the development costs. The winner of this phase would be awarded a contract to develop and build the space shuttle.

McDONNELL DOUGLAS/MARTIN-MARIETTA

The McDonnell Douglas/Martin-Marietta Phase B team also involved TRW Electronics Systems Group, Pan American World Airways, Raytheon, Sperry, Norden, Hamilton Standard and LTV. In a December 1970 report to NASA, this team proposed two different orbiter configurations, and two slightly different derivatives of the same booster to launch them. Both orbiters provided a 15 by 30 ft. payload bay capable of holding 25,000 lbs.

The low cross-range orbiter was a straight wing vehicle spanning 113.8 ft. and weighing 188,600 lbs. empty. Two

415,00 lb. th. boost engines were mounted vertically on the vehicle's centerline similar to the BAC *Lightning*. Four 18,000 lb. th. air-breathing turbofans were mounted in twin pods on each side of the orbiter's mid-fuselage for landing and ferry operations. LH_2 and LO_2 tanks sat end-to-end under the cockpit and payload bay, with the LO_2 tank located forward for better center of gravity control. A secondary LH_2 tank was located behind the payload bay on top of the main LH_2 tank. The fuselage was 147.6 ft. long and 21 ft. wide. The cross-range capability was estimated at 200 miles (substantially less than the *Apollo* capsule was capable of), with a powered landing speed of 178 knots. The wings and fuselage consisted of a titanium heat-sink structure, with the lower surfaces covered with columbium superalloy shingles. Cobalt superalloy was used for the control surfaces, and the vertical tail was a heat-sink structure made of nickel superalloy. Carbon-carbon composite was used on the wing leading edges, and the semi-blended chine was columbium superalloy.

The proposed booster generally was similar to the Martin-Marietta Phase A alternate design. The booster for the straight wing orbiter was 220.2 ft. long with a large aft mounted wing spanning 151 ft. Small canards on the forward fuselage contained ten 18,000 lb. th. turbofan engines for landing operations. Thirteen 415,000 lb. th. rocket engines provided the initial boost to carry the orbiter to staging velocity. The vehicle was projected to have an empty weight of 452,803 lbs. Several designs had been studied for the booster, some with the aft wings mounted on top of the fuselage, some with them mounted low. Also evaluated was the best position of the vertical control surfaces, with some designs having a central vertical tail (as in Phase A) and others having wing-tip mounted tails. The final configuration was a high mounted wing with the outboard tips bent upward to serve as vertical control surfaces. This was found to provide the least interference with the orbiter at launch while still providing adequate stability for the booster. The booster was of aluminum manufacture, with the fuselage composed of integral LH_2 and LO_2 tanks with separate bulkheads, and the wings and canards having conventional spar and rib construction. The main body diameter of 34 ft. was based upon the tooling procured for the *Saturn V* being able to produce cryogenic tanks of that diameter.

The high cross-range orbiter was a sleek blended delta spanning 97.5 ft. and weighing 203,500 lbs. empty. Again, two 415,000 lb. th. boost engines were mounted on the centerline. Four air-breathing turbofans were mounted in pairs in each wing, being deployed down into the airstream when needed during landing. Two LH_2 tanks were located side-by-side under the payload bay, and two LO_2 tanks were contained in the lower forward fuselage. A secondary LH_2 tank was mounted above the LO_2 tanks, but below the cockpit. The total fuselage length was 171 ft. It was projected this design would have a 1,500 mile cross-range capability, and the power-on landing speed was estimated to be 162 knots. The cryogenic propellant tanks were to be of fusion welded aluminum construction, with the rest of the orbiter structure made up primarily of titanium alloys. Heat shields of cobalt superalloy covered the lower wing surfaces, and the wing leading edges were of columbium superalloy. Nickel superalloy shingles covered the sides of the forward fuselage. The vertical tail had a nickel superalloy heat-sink structure and a carbon-carbon leading edge. Some use of boron-epoxy and boron-aluminum was proposed if the technologies matured quickly enough. The booster for the high cross-range orbiter generally was similar though slightly larger and heavier.

Several unmanned second stages also were proposed to supplement the orbiter for heavy lift duties. These second stages included: a modified *Saturn* S-IVB stage with solid rocket motors to augment thrust and capable of carrying 120,000 lbs. into orbit; a modified *Saturn* S-II stage with a nuclear powered upper stage capable of delivering 130,000 lbs.; and an all new second stage built with components derived from the orbiter and booster capable of carrying 170,000 lbs. The total cost per flight of $23.7 million yielded a cost-to-orbit of $198 per pound. The modified S-II/nuclear upper stage would need $106 million in development funds, but would boost a pound into orbit for $186. The all new, shuttle derived, stage would cost $320 million to develop, but would lower the per pound figure to $103. All were considered low risk technology, and the all-new stage was advocated as providing more growth potential and longevity.

The operational baseline for the proposed vehicle included just ten days for turnaround between landing and launch. Three possible launch sites were evaluated: Kennedy Space Center; Vandenberg AFB, California; and the White Sands Missile Range in New Mexico. Since Kennedy was to use existing *Apollo* facilities, it was estimated it would only take five years and $86.75 million to be modified. Vandenberg and White Sands would require new facilities, taking 6 to 7 years to build, and costing $285 million at Vandenberg and $317 million at White Sands. All three sites offered adequate abort options for the high cross-range orbiter, but White Sands and Vandenberg would have no abort options for a first revolution return of the low cross-range version.

NORTH AMERICAN ROCKWELL

The team headed up by North American Rockwell (later to become Rockwell International) included General Dynamics/Convair, IBM, American Airlines and Honeywell. On November 10, 1970 North American Rockwell released their baseline configuration for the Phase B space shuttle competition. The orbiter was a blended delta with the wing tips turned up to act as vertical stabilizers. The overall length was 186.5 ft., and two 415,000 lb. th. LO_2/LH_2 engines were provided. A 15 by 60 ft. payload weighing 25,000 lbs. could be accommodated. The straight winged booster spanned 142 ft. and was 209 ft. 7 in. long. A vee tail was incorporated to minimize the aerodynamic effects of the orbiter upon the vertical control surfaces.

As part of the Phase B study, Rockwell evaluated three different ejection systems: ejection seats similar to the Lockheed seats installed on the A-12/SR-71 series aircraft; encapsulated ejection seats of the type used on the XB-70 experimental bomber; and a separable crew compartment similar in concept to the one used on the F-111 and early B-1A aircraft. The only system that could provide protection for more than the two man flight crew was the separable crew compartment, and that would raise the development costs of the orbiter by almost $300 million and add over 14,000 lbs. to its empty weight (current estimates are that to retrofit an escape module on the existing orbiters would add over 30,000 lbs.). All the systems examined had limitations in their ability to provide successful escape, and all would require advance warning of an impending hazard from reliable data sources.

The Phase B RFP (Para. 1.3.6.2.1) issued during April 1971 states: "Provisions shall be made for rapid emergency egress of the crew during development test flights". The objective was to offer the crew some protection, though limited, from risks during the test flights. The philosophy was that after the test flights, all unknowns would be resolved, and the vehicle would be certified for "operational" use like an airliner. Rockwell eventually selected modifed SR-71 seats for installation in *Enterprise* and *Columbia*. The ejection could be initiated by either crew member and could be used in the event of uncontrolled flight, on-board fire, or pending landing on unprepared surfaces or water. The escape sequence required approximately 15 seconds for the crew to recognize the situation, initiate the ejection sequence, and get a safe distance away from the vehicle.

Although the seats originally were intended for use during first stage ascent, or during gliding flight below 100,000 ft., subsequent analysis shows the crew would be exposed to the solid rocket booster and main engine exhaust plumes if they ejected during ascent. During descent, the seats provided good protection from about 100,000 ft. to landing.

The end result was that a decision was made not to provide a crew escape system for operational shuttle flights. Although this would be highly criticized after the 1986 *Challenger* accident, a more realistic viewpoint was expressed by senior astronaut Robert Crippen: "I don't know of an escape system that would have saved the crew from the particular incident that we just went through (the 51-L accident)."

By the time the Rockwell final report on Phase B was released on March 15, 1972, the idea of a completely reusable two-stage system with a fly-back booster had been abandoned. Rockwell had placed their emphasis on three alternate configurations: a recoverable pressure-fed liquid booster in a series-burn mode; a pump-fed liquid booster based on the F-1 engine from the *Saturn V*; and several configurations using various combinations of solid rocket boosters. Two different orbiter designs were examined, one having a 15 by 60 ft. payload bay, the other having a payload bay measuring 14 by 45 ft. Both orbiters generally resembled the eventual orbiter

NASA

"Pathfinder" during a practice lift in the VAB. Before modification it was called the "Orbiter Simulator".

NASA

Prior to the Boeing 747 carrier aircraft solution, studies were conducted to ascertain the feasibility of transporting Shuttles by dedicated barge. Lengthy transport times between sights eventually killed this concept as an option.

Initially scheduled to be named "Constitution", the first orbiter, OV-101, was renamed "Enterprise" in deference to "Star Trek" television series fans.

"Enterprise" at Rockwell's Palmdale, California final assembly facilty bearing new markings and being prepared for 1983 Paris Airshow.

with the exception of the reaction control sytem being mounted in pods on the tips of the wings and tail instead of co-located with the OMS pods. A parallel-burn configuration for either of the liquid boosters was eliminated because a poor mass fraction resulted in significantly higher costs per flight than the series-burn system. Further cost comparisons between the pressure-fed liquid booster, a 120-in. solid booster, a 156-in. solid, and the *Saturn* based pump-fed liquid booster showed that the liquid feed system had the lowest per flight costs. The 120-in. solid had the highest cost, and was dropped from consideration. This costing analysis showed that 156-in. solids would cost $3.7 billion to develop, with a total system cost of $10.8 billion, resulting in a cost of $228 per pound to orbit. The pressure-fed liquid system would cost $4.6 billion to develop, but the total system would only cost $9.4 billion, with a pound costing $127. The pump-fed liquid booster would put a pound into orbit for $125, after costing $4.2 billion to develop and $8.9 billion to build. It should be noted that the solids, although they had a higher total program cost, would require less early-year funding than the liquid systems, but they would yield higher per-flight costs during operations.

A summary of the technical issues that Rockwell thought were important was also presented. It was felt that the separation dynamics related to the series-burn systems were significantly more straight-forward than those for a parallel-burn system. Experience and relative risk were in favor of the series-burn systems because of the long history of successful staged launch vehicles. The parallel-burn system was thought to have some advantages in ground handling, and since the SSMEs were started prior to liftoff it gave some assurance that they were indeed startable, and running stable. It was felt that the liquid propellant system was more flexible because of the ability to tailor the thrust at almost any point in the ascent phase, however, the development risk appeared to be in favor of the solids because of their greater simplicity and the fact that recovery of the liquid boosters presented more of a challenge, since they were more fragile. Ground handling of the solid boosters also was considered easier. On the bottom line, Rockwell recommended a series-burn (staged) system utilizing solid propellant boosters.

ALTERNATE CONCEPTS

While the Phase 'A' and 'B' studies were being directed by the Manned Spacecraft Center in Houston, the Marshall Space Flight Center in Huntsville sought a second opinion. The study of alternate space shuttle concepts (ASSC) was initiated on July 6, 1970 in parallel with the Phase B studies of the fully reusable two-stage-to-orbit shuttle configurations. The intent of the ASSC study was to define and investigate various promising alternatives to the fully reusable concepts so that meaningful comparisons of technological issues, operational approaches and cost factors could be made. It also allowed NASA a fallback position if space shuttle funding was reduced. Contracts were issued to Lockheed, Chrysler and a Grumman/Boeing team. The initial matrix of alternate concepts to be studied consisted of 29 configurations in three general categories: "fractional" stages (stage-and-a-half); semi-reusable (expendable booster/reusable orbiter); and two-stage fully reusable alternatives similar to the Phase B vehicles. A fourth general category subsequently was added—reusable orbiters having expendable external fuel tanks.

Ultimately the studies concluded that the fully reusable two-stage-to-orbit vehicles being developed under Phase B were the "best" since they gave the greatest assurance of achieving low cost per flight into orbit, and would have the lowest "total" program costs. However if peak funding and/or development risk was to be minimized, consideration should be given to a phased development approach of a reusable orbiter with an expendable booster (probably based on *Saturn's* S-1C stage) with the option of developing a recoverable booster later.

During May 1971, OMB officials told NASA that it could expect to get no budget increases in the next five years. This was a drastic blow because it meant the agency could not carry out the shuttle development program it had been planning for almost two years during the ILRV studies. Those plans called for a development cost of almost $10 billion with a peak annual budget of some $2 billion. If limited to the $3.2 billion budget just approved for FY72, the most NASA could hope to put into shuttle development, and still maintain a balanced science and application program, was roughly $1 billion a year for five years.

In addition to the ASSC contracts, several contractors expended company R&D funds to investigate other possible design concepts. As a result of these expenditures, Grumman, during the fall of 1970, developed the concept of moving the liquid hydrogen tanks outside the orbiter fuselage and jettisoning them when empty. This would have a profound influence on the course of space shuttle development.

GRUMMAN/BOEING

On December 23, 1970 the Grumman/Boeing ASSC team received NASA approval to conduct parallel studies of reusable two-stage configurations employing internally and externally mounted orbiter liquid hydrogen tanks. Reviews during January and March 1971 indicated significant weight, development and cost advantages for the external hydrogen tank orbiter/heat-sink booster concept. During a subsequent study effort, both two- and three-engined orbiters embodying the external hydrogen tank/heat-sink booster design were compared to a representative internal hydrogen tank configuration. The results of these studies were reviewed by NASA during April 1971 and as a result, Grumman/Boeing were authorized to study a three-engined external hydrogen tank/heat-sink booster vehicle utilizing 415,000 lb. th. rocket engines.

The vehicle that emerged included a 157 ft. long orbiter powered by three main engines of 415,000 lbs. th. A blended 55° sweep delta wing spanned 97 ft., and the sweptback tail was 61.25 ft. high. Four air-breathing JTF-22A4 landing engines were provided, and a cross-range capability of 1,100 miles was projected. A landing speed of 180 knots was anticipated, and a ferry range of 300 miles could be achieved using the air-breathers. The orbiter had an empty weight of 197,000 lbs. Two internal uninsulated LO_2 tanks held 705,696 lbs. of oxidizer, and two external LH_2 tanks each carried 59,537 lbs. of fuel each. The booster was 173 ft. long, weighed 220,135 lbs. empty and employed a straight wing spanning 98.4 ft. Twelve rocket engines of the same type as the orbiter's provided assent boost, and eight F101 jet engines allowed the booster to fly back to its launch site. A ferry range of 443 miles could be obtained. Thermal protection during reentry was provided by the heat capacitance of the basic structure, hence the term "heat-sink". This feature of a single shell structure reacting to both flight

"Enterprise" during approach to Edwards AFB on its fifth and final free flight in the 13-flight ALT program. Split rudder-type speed brake is noteworthy.

loads and heat loads significantly reduced booster weight. The trade-off was that the orbiter was launched at a lower staging velocity, and thus required more propellants. However, this too was offset by moving the LH_2 tanks outside the airframe.

It was anticipated that with a development go-ahead during March 1972, the first flight test orbiter would be completed during June 1973, with the first booster following during April 1976. The first horizontal flight tests would be conducted during April 1978, with a fully operational system in place by July 1979. At peak traffic rates for a mature system, it was estimated that 650 people could support direct operations, with the total rising to 1,000 when engineering support, quality control and other technical support personnel were included. If all other support personnel (fire, mail, janitorial, etc.) were included, the total rose to about 3,000 peopel to support 75 launches per year. This was an average cost (1970 dollars) per launch for manpower of $540,000.

LOCKHEED

The Lockheed Missiles and Space Company's study of Alternate Space Shuttle Concepts was released on June 4, 1971. The concept that Lockheed expended roughly 75% of its efforts on was a "stage-and-one-half" vehicle that later could be converted for use in a true two-stage reusable shuttle system.

The stage-and-one-half vehicle consisted of a 156.5 ft. long, delta-body orbiter with a large, V-shaped drop tank. The orbiter contained all main propulsion systems, some ascent propellants, the payload, the crew, and all supporting systems. The expendable tanks, 192 ft. long, with a diameter of 27 ft., contained most of the ascent propellant which was cross-fed into the orbiter main engines. There were no systems within the drop tanks except for propellant instrumentation, and external insulation.

The perceived advantages to the stage-and-one-half concept were:

- Only one complex vehicle (the orbiter) needed developing. This would result in lower development costs, low total program costs, lower annual peak funding, and low sensitivity of program costs to vehicle changes.
- Only one flight test program, crew training program, etc. would be required.
- All engines would be ignited and checked out prior to lift-off.
- Safety would be enhanced by virtue of there being only one crew and one vehicle.
- No advanced aerodynamic parallel staging technology development was required.
- The boost configuration was relatively "clean", and thereby provided low risk of major aeroelastic and dynamic development problems.
- The system could be converted to a true two-stage system, allowing a possible reduction in recurring costs.

The only significant advantages the two-stage vehicle had over the stage-and-one-half concept were: lower recurring costs due to the use of all fully reusable elements; and, no launch azimuth restraints as long as both stages had an intact abort capability. Lockheed felt that developing the stage-and-one-half vehicle would lower all development risks, and their associated costs, thereby allowing NASA to build a usable shuttle within the available budget, and at the same time reserve the option for a fully reusable vehicle when funding could be found to develop the booster.

The orbiter was a delta-body vehicle having high volumetric efficiency, large leading edge radii, and a steeply sloped upper forward body to provide satisfactory pilot visibility. Comparison of the delta-body orbiter to a wing-body vehicle showed that the delta-body was lighter by 20,000 lbs., and would save $500 million in total program costs. It represented a lower aerothermodynamic development risk and could accomplish over 1,500 nm of cross-range. The sloped vertical fins provided directional stability, longitudinal stability and increased subsonic and hypersonic lift. Avionics were to be located in the extreme nose, followed by the crew compartment which was joined to the 15 by 60 ft. payload bay by a crew transfer tunnel. Payload erection and deployment equipment was located at the forward end of the payload bay. Six Pratt & Whitney JTF22-A4 air-breathing engines were located in the bottom of the vehicle, and deployed downward for subsonic landing, go-around and ferry missions. Nine 530,000 lb. th. rocket engines were installed at the extreme rear of the vehicle. The orbiter airframe structure was a conventional aluminum stringer and bulkhead type attached to the

NASA

"Enterprise" being removed from dynamics test stand at the MSFC during August 1978.

thrust structure. It could be disassembled for complete removal and replacement of the non-integral internal propellant tanks. The internal propellant tanks were arranged in two sets, with the liquid hydrogen tank forward and liquid oxygen aft. The titanium thrust structure transmitted the main engine thrust loads directly to the orbiter propellant tanks, the drop tanks, and the aft orbiter airframe shell and fins. The upper surface thermal protection system was to be titanium for temperatures up to 1,000° F. The lower surfaces, fin surfaces, base heat shield, and leading edges would be protected with LI-1500 silica tiles. The nose cap was to be a coated tantalum and silicone system capable of sustaining temperatures of 3,000° F, or more. Total gross take-off weight was estimated at 3,816,420 lbs.

Lockheed anticipated using the Launch Umbilical Towers used during the *Apollo* project, with minor modifications, to service the orbiter on the launch pad. All nine engines were ignited on the pad at liftoff. Staging occurred at 18,000 ft. per second velocity, near zero aerodynamic pressure (zero "q"), with only three engines burning. The external tank was separated by releasing the forward attach pin, and letting aerodynamic forces carry the front of the tank up and over the orbiter, pivoting on the rear attach points. When the tank was roughly vertical to the orbiter, the rear attach points (which also contained the propellant feed lines) would be released, and the orbiter pitched slightly downward, allowing the tank to pass over the orbiter and burn up in the atmosphere. Total tank separation time was roughly 10 seconds. The

NASA

"Enterprise" being rolled out of the VAB at 8:24 a.m. EDT, May 1, 1979, for the 3-1/2 mile journey to Pad A.

NASA

Stand was modified prior to 1978 Mated Vertical Ground Vibration Tests.

orbiter continued into orbit using internal propellants until sufficient velocity was achieved to place it into the desired orbit.

The directions of the Alternate Space Shuttle Concepts studies indicated the contractors were not to do detail subsystem design unless the subsystem was unique to the alternate concept. Therefore, Lockheed did not spend much effort to determine the booster configuration that would be used to launch the orbiter if it was upgraded to a true two-stage system. The booster used in the study was a scaled up version of the Phase 'B' configuration being proposed by McDonnell Douglas/Martin-Marietta. It was expected that the orbiter would be attached in the same location as the Phase 'B' orbiter. For use in the two-stage system, the Lockheed orbiter would not carry its external drop tanks, and seven of the nine rocket and two air-breathing engines would be removed. Internal propellant tankage would be increased to feed the remaining engines during second stage ascent, and the configuration of the internal tanks would be reversed to assist in controlling c.g. shifts. The orbiter would carry a 11,500 lb. weight penalty imposed by a thrust structure designed to handle nine main engines, instead of the two required for a two-stage system. It was anticipated this would be overcome by redesigning this one component on any new-build orbiters.

CHRYSLER

Chrysler proposed what surely was the most unique

NASA

The May 1 move was completed at 7:20 p.m. when the MLP was hard-down on its mounts at Pad A.

NASA via Erick Simonsen

"Enterprise" landing during the ALT flight series conducted at Edwards AFB. Trailing-edge-up angle of control surfaces, and split rudder are noteworthy.

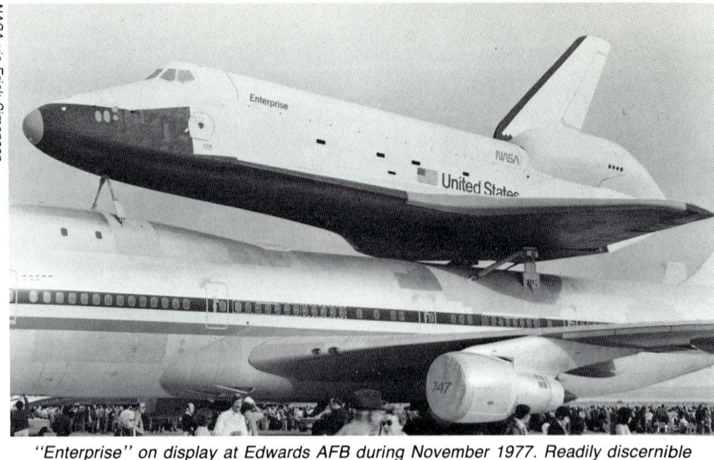

Gerald Balzer

"Enterprise" on display at Edwards AFB during November 1977. Readily discernible are special mounting assemblies for attachment to Boeing 747-123 carrier aircraft.

Dennis Jenkins

"Enterprise" and Boeing 747-123 taking off for first ALT flight. Visible on fuselage side of N-905NA is barely discernible "American Airlines" logo.

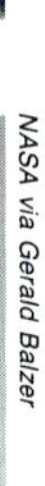

Post-flight view of "Enterprise" following October 12, 1977 ALT FF-4 at Edwards AFB. This was the first "tail cone off" free flight.

alternate shuttle concept. Looking like an overgrown *Apollo* command module, SERV (Single-stage Earth-orbital Reusable Vehicle) had a diameter of 90 ft. and was 66.5 ft. high. A payload bay measuring 23 ft. in diameter and 60 ft. in length could hold 88,060 lbs. destined for a 100 NM orbit at 55° inclination. A 12-module LO_2/LH_2 integral aerospike engine was 88.7 ft. in diameter and 8.2 ft. long, developing 5.4 million lbs. th. and a specific impulse of 346.7 seconds at sea level. Total lift-off weight was 4.5 million pounds. Twenty-eight JP-4 fueled turbojet lift engines provided deceleration and landing propulsion. This booster could be launched unmanned into orbit to be unloaded by an already orbiting "space-tug", or alternately, could carry a manned winged spacecraft resembling the *Dyna-Soar* on top.

Upon command to return to earth, SERV reoriented itself into the familiar *Apollo* command module reentry attitude, and entered the atmosphere semi-ballistically. With SERV, however, the target landing point was designed to be land instead of water. In fact, it was to be a landing pad within a few hundred yards of the Vehicle Assembly Building (VAB) at the Kennedy Space Center. Using existing *Apollo* reentry navigation and guidance technology, it was projected SERV would be within four miles of the landing pad when it reached 25,000 ft. At that altitude, air intakes and doors for exhaust efflux would open, and the four groups of seven air-breathing lift enignes would ignite, providing SERV with a soft landing.

It was proposed that construction of SERV would be at the Michoud Assembly Facility (MAF) used to construct the *Saturn V* first stage (and, currently, the ET). The vehicle then would be transported by a *Bay*-class vessel owned by the West India Shipping Company that was 266 ft. long with a 51 ft. beam and a six foot draft. "Pin-on" type sponsons would increase the beam to 89 ft., enabling it to carry SERV and still pass through the locks and bridges between Michoud and KSC with no route modifications.

Plans called for a 12 month preliminary design effort, followed by ten months of dtailed design. One test article would be constructed, and 20 months of static testing would be performed. A single propulsion system test article would also be tested concurrently. Three flight test vehicles then would be procured. Total development time would be four years, and cost $3.565 billion. Each production flight vehicle would cost $350 million in 1971 dollars, and have a 10 year/100 flight life.

REDIRECTION

Most ideas for lowering the costs of the space shuttle had involved replacing the flyback booster with some sort of expendable stage. This did not appear technically feasible, however, given the large size of the orbiter. This problem was resolved during June 1971 with a NASA decision to endorse the concept developed by Grumman to move the large liquid hydrogen tanks outside the airframe, and make them expendable. This made the orbiter smaller and lighter, with a significant reduction in development costs, but with a corollary increase in costs each time the orbiter was launched. Further study showed it would be even more cost effective to place both the liquid hydrogen and liquid oxygen tanks externally in a single, expendable structure. On September 12, 1971, the Phase B contractors were told to re-design the orbiter using the "external tank" concept.

During August 1971, the Spacecraft Design Directorate at the Manned Spacecraft Center had derived the "040" configuration based on the results presented so far from the Phase B studies. This clipped-delta wing, three engine orbiter using an external tank would become the baseline for evaluating further Phase B proposals. During this same period, a decision already had been made to remove the air-breathing landing engines after the first couple of years of flight to recover the 20,000 lb. (engines, mounts and fuel) payload penalty. NASA also had to choose between two options for providing an ascent abort capability for the orbiter. In one case Abort Solid Rocket Motors (ASRMs) were mounted on the aft portion of the orbiter fuselage, above the wing but below the OMS pods. This solution was deemed appropriate for either a series-burn (staged) or parallel-burn configuration. The second option incorporated the ASRMs on the ET interstage between the liquid oxygen and liquid hydrogen tanks. This option was only appropriate for a series-burn system. It was determined that the weight penalty imposed by the ASRM system could be offset by the performance gained through the use of the abort system during a nominal ascent. Specifically, when the vehicle had achieved enough velocity to reach low-orbit even without the abort motors, they would be fired to provide the additional thrust needed to reach a higher orbit. Careful sequencing of the ASRM firing was required together with throttling of the orbiter main propulsion system to avoid over-accelerating the orbiter. Through this technique, less liquid propellants (LO_2/LH_2) could be carried, recovering most of the weight penalty imposed by the ASRM system. The abort motors were roughly 30 ft. long and 5 ft. in diameter. NASA's rationale for requiring the ASRM system was that in order to keep the main rocket boosters simple and inexpensive, no thrust vector control system would be incorporated on them. This would require another method of ensuring the orbiter could abort early in the ascent phase, hence the ASRMs.

As late as the end of November 1971, NASA was still undecided on whether the orbiter's engines should be ignited simultaneously with the booster's (the "parallel burn" approach), or if a more traditional "staged" approach of letting the booster power the initial ascent, then ignite the orbiter's engines for the final climb to orbit should be taken, although it was tending toward the "parallel burn" design. Mathematica, the consulting firm conducting the shuttle economic analysis for NASA, reported that a parallel burn shuttle with an external fuel tank was the economically preferred concept, and that it could be developed for approximately $6 billion, with a cost per launch of around six million dollars. During December 1971 NASA adopted the parallel burn design, but left open the decision of liquid or solid fueled boosters. By this time, the MSC-040 design had progressed to the "040C" variant, giving up its clipped-delta wing for a somewhat blended-delta. Total development costs now were estimated at $5.8 billion. The booster decision finally was announced on March 15, 1972, with the choice being solids. Fletcher reported to the OMB that solid boosters could be developed faster and for $700 million less than an equivalent thrust liquid booster lowering total development costs to $5.15 billion in 1971 dollars. However, a solid booster would cost more to operate, and contained more unknowns, since NASA had little experience with this technology. This decision placed a large share of the burden of paying for the shuttle program on its future users. The 1972 mission model for NASA, DoD and other users called for some 580 flights over a 12 year period (1979-1990), an average of

NASA

Three of the CRTs and two of the keyboards that provide the interface to the General Purpose Computers are shown in this shot of the flight deck of "Enterprise". Additional CRTs and keyboards are located in the aft cabin for use by the payload specialists and while working at the remote manipulator arm control station.

NASA

"Columbia" at Launch Complex 39. Readily discernible were black wing chines peculiar to this vehicle. Early wing markings also are prominent; later orbiters carried the U.S. flag and "USA" on the left wing and the orbiter's name and "NASA" on the right. Large pole on top of launch tower serves as lightning rod.

Rockwell International Space Shuttle, OV-099, "Challenger", as seen in tile patterns consistent for most of its missions. This pattern was the same as used on "Challenger" during the fateful STS-33 (51-L) mission. The nose radome and wing leading edge were in standard non-gloss light gray-green coated reinforced carbon-carbon (RCC); undersurfaces, portions of nose/cockpit sections and vertical fin/rudder were in black HRSI tiles. The balance of the spacecraft was in white LRSI tiles and white FRSI blanket insulation.

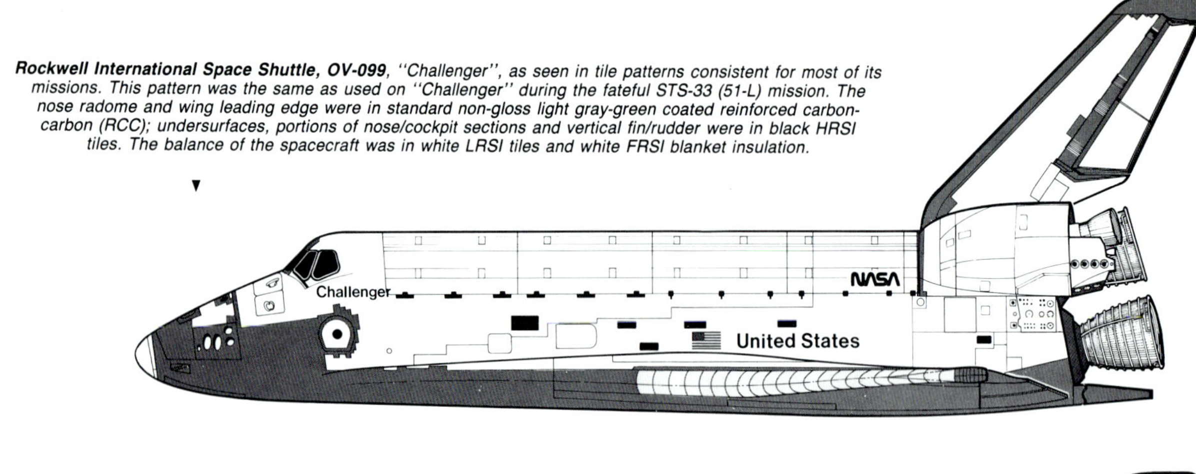

Rockwell International Space Shuttle, OV-104, "Atlantis", as it appeared during first two flights. HRSI tile pattern and RCC installation was similar to that used on early OVs. "Atlantis", the newest Orbiter Vehicle, is the lightest weight OV yet produced because of the extensive use of FRSI blanket insulation.

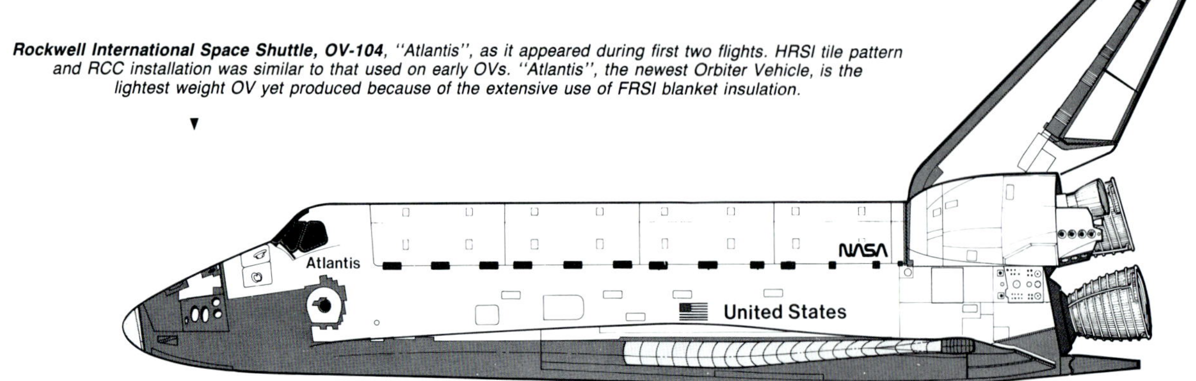

Shuttle Derived Vehicle (SDV), Class I baseline configuration, also known as "Shuttle-C". The vehicle is designed for a 25 ft. dia. x 90 ft. long payload of 134,536 lbs. Three standard SSME shuttle engines and two OMS type maneuvering rockets make up the powerplant designed for this vehicle. Attachment of the SDV Class I to standard ET/SRB vehicle is accomplished without revision to the standard mounts.

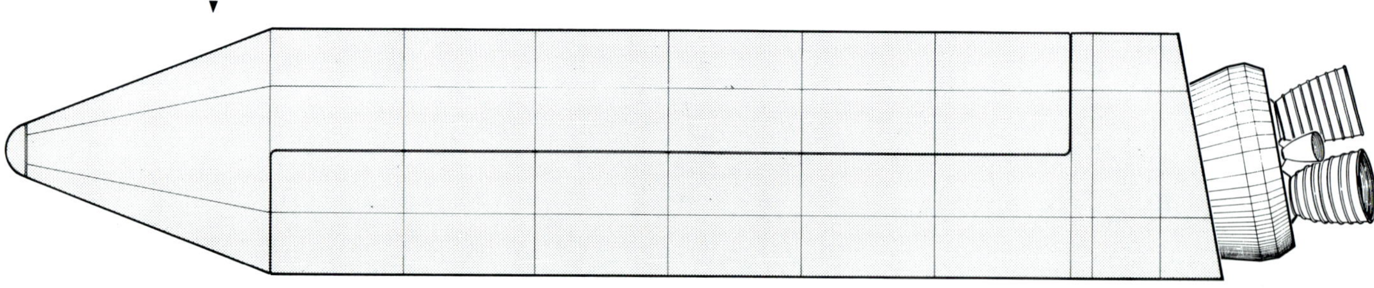

Rockwell International Space Shuttle, OV-103, "Discovery", in standard HRSI, RCC, and extensive FRSI patterns. "Discovery" is shown with its non-retractable landing gear extended. Since the space shuttle series is not equipped for a powered missed-approach or go-around, a weight savings was achieved through the deletion of retract mechanisms.

Scale: 1/250th

Drawn by Mike Wagnon

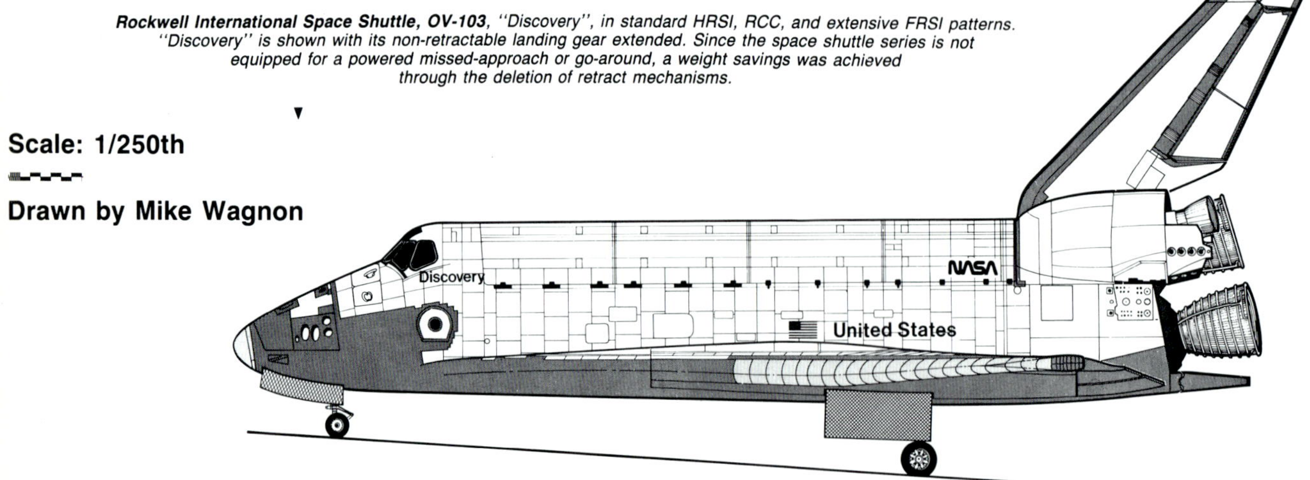

SELECT MARKINGS

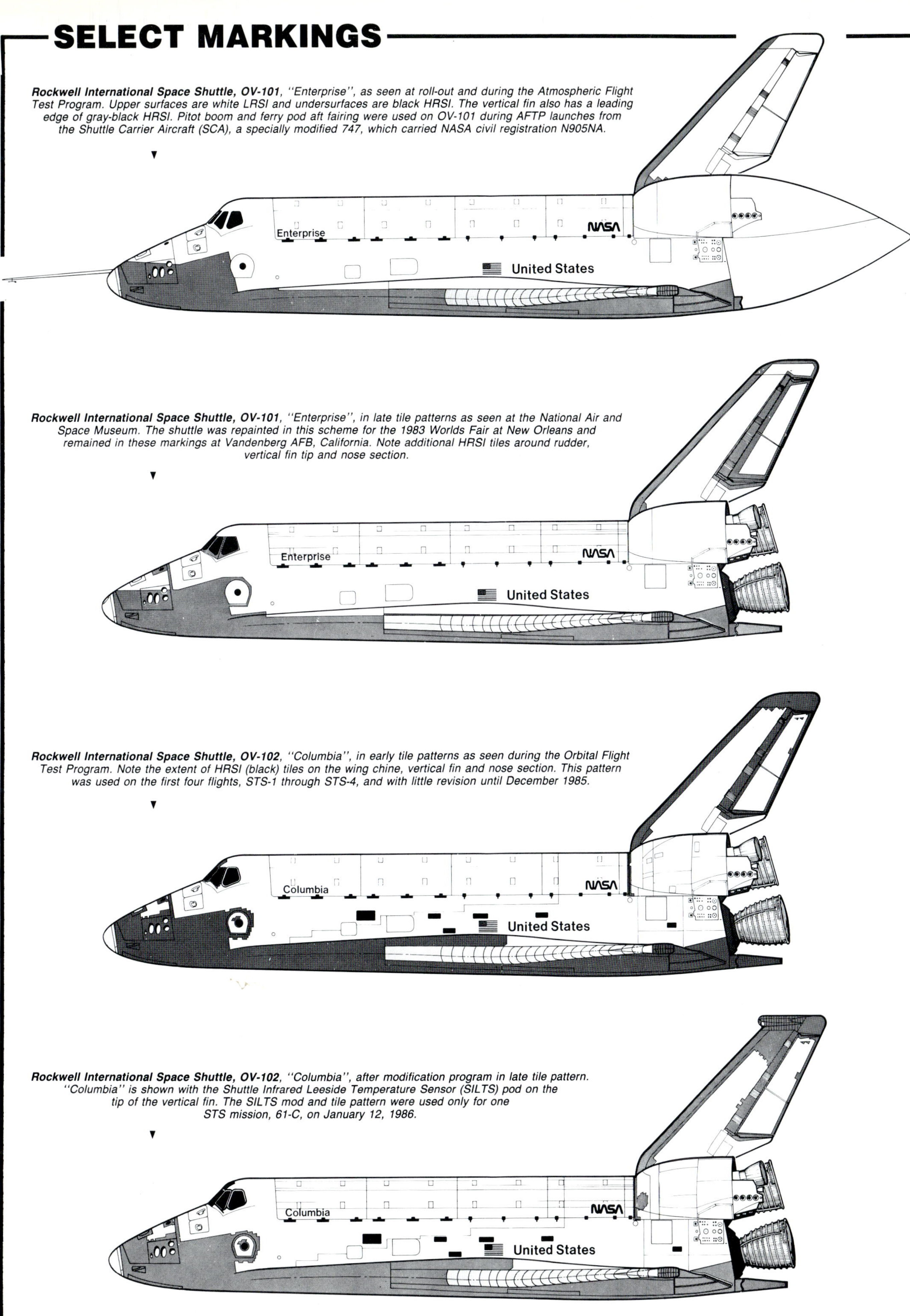

***Rockwell International Space Shuttle, OV-101**, "Enterprise", as seen at roll-out and during the Atmospheric Flight Test Program. Upper surfaces are white LRSI and undersurfaces are black HRSI. The vertical fin also has a leading edge of gray-black HRSI. Pitot boom and ferry pod aft fairing were used on OV-101 during AFTP launches from the Shuttle Carrier Aircraft (SCA), a specially modified 747, which carried NASA civil registration N905NA.*

***Rockwell International Space Shuttle, OV-101**, "Enterprise", in late tile patterns as seen at the National Air and Space Museum. The shuttle was repainted in this scheme for the 1983 Worlds Fair at New Orleans and remained in these markings at Vandenberg AFB, California. Note additional HRSI tiles around rudder, vertical fin tip and nose section.*

***Rockwell International Space Shuttle, OV-102**, "Columbia", in early tile patterns as seen during the Orbital Flight Test Program. Note the extent of HRSI (black) tiles on the wing chine, vertical fin and nose section. This pattern was used on the first four flights, STS-1 through STS-4, and with little revision until December 1985.*

***Rockwell International Space Shuttle, OV-102**, "Columbia", after modification program in late tile pattern. "Columbia" is shown with the Shuttle Infrared Leeside Temperature Sensor (SILTS) pod on the tip of the vertical fin. The SILTS mod and tile pattern were used only for one STS mission, 61-C, on January 12, 1986.*

NASA

"Challenger" at the beginning of the 3.5 mile trek to the Launch Complex. The Mobile Launch Platform (MLP) on which the orbiter sits weighs 8,230,000 lbs. and is carried by a Crawler-Transporter that weighs 6,300,000 lbs. The latter vehicle is capable of a top speed of one mile per hour during orbiter delivery.

NASA via Erik Simonsen

''Columbia'' at the beginning of STS-2. At the bottom is sound-suppressing spray water entering the main engine cut-out in the MLP. The corrugated ET section is known as the ''inter-tank'' and covers the space between the liquid oxygen and hydrogen tanks. SRB attachment point load bearing structure also is located here.

Rockwell International via Erik Simonsen

"Columbia" exiting the VAB in preparation for STS-2. This was the last flight to use a painted ET. Shuttle assemblies usually are rolled out just after midnight as this historically is the time of least lightning activity in central Florida. Size of VAB is readily apparent in this view as shuttle assembly alone stands over 18 stories.

ROCKWELL INTERNATIONAL SPACE SHUTTLE, OV-103, *DISCOVERY*

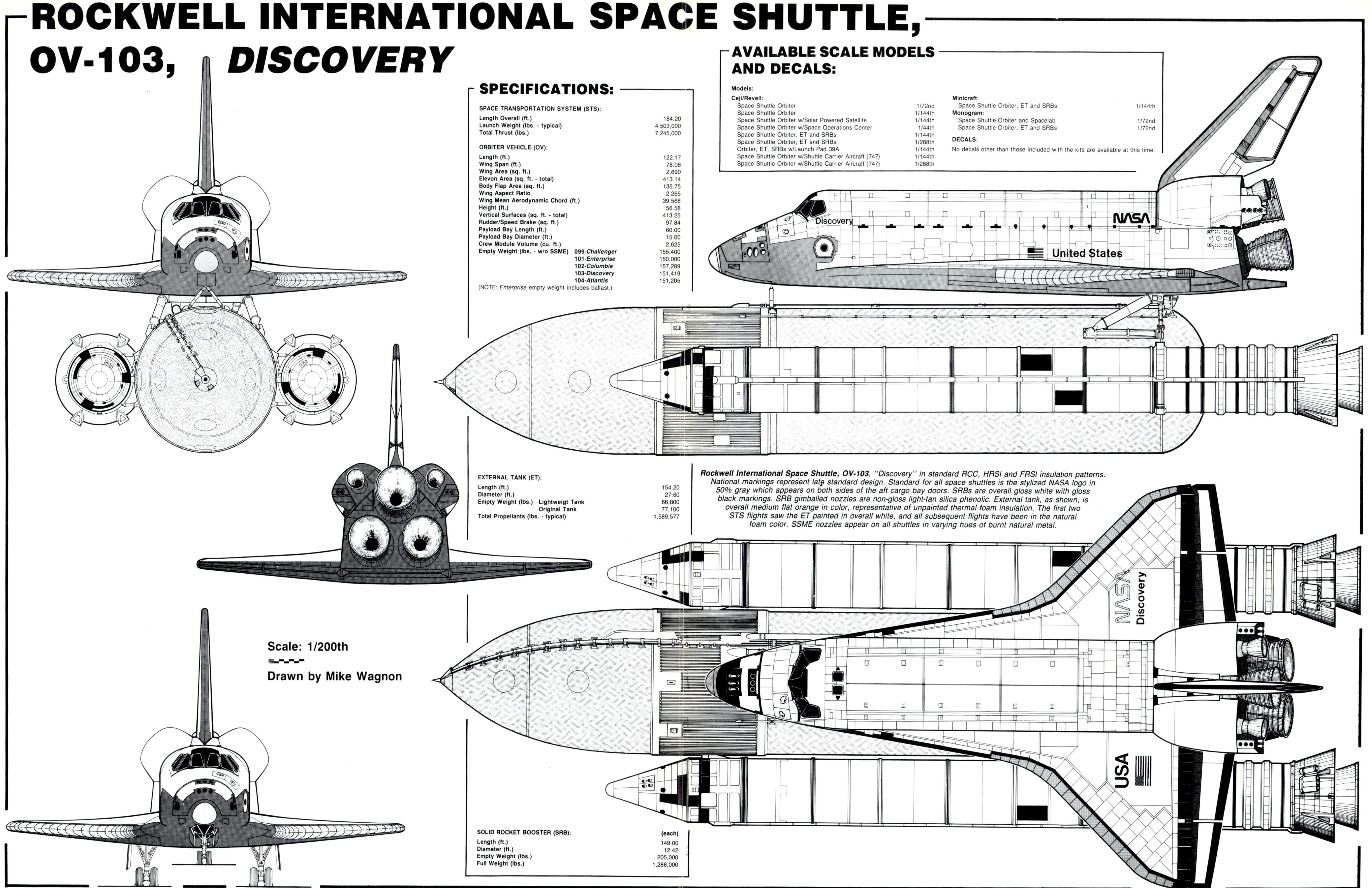

SPECIFICATIONS:

SPACE TRANSPORTATION SYSTEM (STS):

Length Overall (ft.)	184.20
Launch Weight (lbs. - typical)	4,503,000
Total Thrust (lbs.)	7,245,000

ORBITER VEHICLE (OV):

Length (ft.)		122.17
Wing Span (ft.)		78.06
Wing Area (sq. ft.)		2,690
Elevon Area (sq. ft. - total)		413.14
Body Flap Area (sq. ft.)		135.75
Wing Aspect Ratio		2.265
Wing Mean Aerodynamic Chord (ft.)		39.568
Height (ft.)		56.58
Vertical Surfaces (sq. ft. - total)		413.25
Rudder/Speed Brake (sq. ft.)		97.84
Payload Bay Length (ft.)		60.00
Payload Bay Diameter (ft.)		15.00
Crew Module Volume (cu. ft.)		2,625
Empty Weight (lbs. - w/o SSME)	099-*Challenger*	155,400
	101-*Enterprise*	150,000
	102-*Columbia*	157,289
	103-*Discovery*	151,419
	104-*Atlantis*	151,205

(NOTE: *Enterprise* empty weight includes ballast.)

EXTERNAL TANK (ET):

Length (ft.)		154.20
Diameter (ft.)		27.60
Empty Weight (lbs.)	Lightweigt Tank	66,800
	Original Tank	77,100
Total Propellants (lbs. - typical)		1,589,577

SOLID ROCKET BOOSTER (SRB):

	(each)
Length (ft.)	149.00
Diameter (ft.)	12.42
Empty Weight (lbs.)	205,000
Full Weight (lbs.)	1,286,000

AVAILABLE SCALE MODELS AND DECALS:

Models:

Ceji/Revell:

Space Shuttle Orbiter	1/72nd
Space Shuttle Orbiter	1/144th
Space Shuttle Orbiter w/Solar Powered Satellite	1/144th
Space Shuttle Orbiter w/Space Operations Center	1/44th
Space Shuttle Orbiter, ET and SRBs	1/144th
Space Shuttle Orbiter, ET and SRBs	1/288th
Orbiter, ET, SRBs w/Launch Pad 39A	1/144th
Space Shuttle Orbiter w/Shuttle Carrier Aircraft (747)	1/144th
Space Shuttle Orbiter w/Shuttle Carrier Aircraft (747)	1/288th

Minicraft:

Space Shuttle Orbiter, ET and SRBs	1/144th

Monogram:

Space Shuttle Orbiter and Spacelab	1/72nd
Space Shuttle Orbiter, ET and SRBs	1/72nd

DECALS:

No decals other than those included with the kits are available at this time.

Rockwell International Space Shuttle, OV-103, *"Discovery" in standard RCC, HRSI and FRSI insulation patterns. National markings represent late standard design. Standard for all space shuttles is the stylized NASA logo in 50% gray which appears on both sides of the aft cargo bay doors. SRBs are overall gloss white with gloss black markings. SRB gimballed nozzles are non-gloss light-tan silica phenolic. External tank, as shown, is overall medium flat orange in color, representative of unpainted thermal foam insulation. The first two STS flights saw the ET painted in overall white, and all subsequent flights have been in the natural foam color. SSME nozzles appear on all shuttles in varying hues of burnt natural metal.*

Scale: 1/200th

Drawn by Mike Wagnon

NASA via Erik Simonsen

"Columbia" at the beginning of STS-2. At the bottom is sound-suppressing spray water entering the main engine cut-out in the MLP. The corrugated ET section is known as the "inter-tank" and covers the space between the liquid oxygen and hydrogen tanks. SRB attachment point load bearing structure also is located here.

Rockwell International via Erik Simonsen

"Columbia" exiting the VAB in preparation for STS-2. This was the last flight to use a painted ET. Shuttle assemblies usually are rolled out just after midnight as this historically is the time of least lightning activity in central Florida. Size of VAB is readily apparent in this view as shuttle assembly alone stands over 18 stories.

less than 50 flights per year. Thirty-eight percent of these flights were to be scientific research aboard the Spacelabs modules; 31% were to be in support of the DoD; and the remaining 31% were to be commercial satellite deliveries and space station support missions. Later (1974) models would call for over 60 flights per year, and as late as 1985, just prior to the *Challenger* accident, NASA still was predicting 20 flights per year. Using the 1972 model, launch and launch-related costs using existing expendable vehicles were estimated at $13.2 billion, while the total shuttle launch costs were estimated at $8.1 billion excluding research and development costs. The cost per shuttle flight was estimated at $10.5 million, or $175 per pound for a full payload bay. (During 1988, the OMB found it cost $300 million per flight, or $5,000 per pound and that development had cost $9.912 billion in 1982 dollars, or $6.651 billion during 1971 dollars—$1.5 billion more than estimated.) It was expected the shuttle system could be brought to the pad and checked-out, then placed on "stand-by" status for 24 hours at a time. An orbiter on stand-by would require only two hours notice for cryogenic fueling (LO_2/LH_2) and crew ingress prior to launch. This was a capability highly valued by the DoD, but one that would prove to be unobtainable in actual operations.

CONTRACT AWARD

On July 26, 1972 Rockwell International won the $2.6 billion contract to design and build the space shuttle orbiter, mandated to fly 100 times each, and capable of 60 flights per year. The design was to be based on the final MSC-040C configuration. Due east flight from the Kennedy Space Center would carry up to 65,000 lbs. in a payload bay 15 x 60 ft. in size and the vehicle would be capable of a 1,085 mile reentry maneuver on either side of its flight path. The contract also included the systems integration function, where Rockwell would ensure all elements of the space shuttle system worked together. Martin-Marietta was selected on August 16, 1973 to design, develop and test the 154 ft. long external tank (ET). This initial contract included three ground test tanks and six development flight tanks. Morton-Thiokol was selected to fabricate the Solid Rocket Boosters (SRB); 149 ft. long and 156 in. in diameter, producing 2.65 million pounds of thrust for two and a half minutes, they would be the largest most powerful solid boosters ever built. The Rocketdyne Division of Rockwell already had been contracted during July 1971 to build the Space Shuttle Main Engines (SSMEs).

By early 1973 several design changes had been approved. Continued studies of abort profiles indicated the thrust termination system planned for the SRBs could be dropped. The thrust termination system was designed to extinguish or reduce the thrust of the SRBs in an emergency situation. With the thrust terminated, additional options would be available for aborts or crew escape. The principle drawback was that the thrust termination occurred so suddenly that it introduced dynamic loads that could cause orbiter structural components to fail. Design reviews and analysis showed that to design the orbiter to withstand the stresses caused by rapid thrust termination would add an additional, prohibitive 19,600 lbs. to the empty weight of the orbiter, so the thrust termination requirement was dropped on April 27, 1973. The Abort Solid Rocket Motors also were deleted as a cost and weight saving measure since it was estimated the ASRMs would cost over $300 million to design and build, and could provide meaningful assistance during just 30 seconds of each flight. A decision was made to equip the orbiter with a parachute braking system during flight tests to reduce its landing roll-out. By November 1973, the air-breathing engines had been deleted as a permanent installation. In their place would be the Air-Breathing Propulsion System (ABPS) consisting of five TF33-P7A turbofans in three pods, a two-engine pod under each wing, and a single-engine pod under the centerline, for use during ferry flights. Fuel for the engines would be carried in a special tank installed in the payload bay. At this time, Rockwell's design had a forward reaction control system that was contained behind doors that opened after the orbiter reached orbit. Further study indicated it would be significantly less complex (and less expensive) to design a system capable of withstanding the rigors of ascent and reentry while exposed, leading to the system in use on the current orbiters.

As 1974 arrived, fabrication of the first vehicle began. As construction continued, NASA deleted the parachute braking system as unnecessary: Edward's lakebeds were more than long enough. The delta-wing planform was modified to form a blended double-delta, and the leading edge sweep was changed from 50° to 40°. The fuselage was shortened by removing the permanent docking attachment, placing a removable one in the payload bay when it was needed, at a sacrifice of some payload capacity. The OMS pods next to the vertical fin were redesigned not to overlap the payload bay doors. This simplified the pod/door interface, and allowed the use of a one-piece (per side) door instead of the four-piece door originally envisioned. The hydraulic system was redesigned to incorporate three redundant systems, instead of the earlier four. The last major configuration change was the final deletion of any air-breathing engine capability in an attempt to get the orbiter weight under 150,000 lbs. and reduce complexity.

SHUTTLE DERIVED VEHICLES (SDV)

Almost as soon as the contracts were awarded to Rockwell to build the orbiters, NASA started seeking ways of improving the payload capability of the shuttle system. Various studies eventually were undertaken by several contractors calling for shuttle modifications, or building new vehicles using shuttle designed components and launch facilities. One such study was completed by Martin-Marietta during May 1982. It identified four possible shuttle derived vehicles, as follows:

- Class I: The orbiter is replaced with a Cargo Element, and a reusable Propulsion/Avionics (P/A) package.
- Class II: The solid rocket boosters (SRBs) are replaced with reusable liquid rocket boosters (LRB) of higher thrust.
- Class III: The orbiter is replaced by the Cargo Element and P/A package, and the SRBs are replaced with the LRBs.
- Class IV: The propulsion system and appropriate avionics are removed from the orbiter and installed under the external tank (ET).

Rockwell International via Erik Simonsen

"Enterprise" at Edwards AFB being prepared for mating to carrier aircraft. It is seen following a repainting for display at the 1983 Paris Airshow.

Dennis Jenkins

Lightweight, transportable "lifting frame" being demonstrated and tested at the Vandenberg Launch Site. Both vehicles had been repainted for 1983 Paris Airshow.

NASA via Erik Simonsen

"Enterprise" during final ALT flight and while approaching Edwards AFB runway 22. This was the second flight conducted with the "tail cone off".

NASA via Dennis Jenkins

"Enterprise" on display at the New Orleans World's Fair during 1984. It was delivered to the site by barge—a first and last for the shuttle series.

NASA via Gerald Balzer

"Columbia" after returning from STS-1. Air-conditioning and purge umbilicals already have been attached. Each orbiter mission ends with a "sniffer" team checking the vehicle for leaks from the hypergolic fuel and oxidizer systems. After the area has been declared safe, other post-flight activity is initiated.

NASA via Gerald Balzer

"Challenger" touching down at Edwards AFB on November 6, 1985, at the end of STS-30 (61-A). The effects of reentry heating are visible on the forward fuselage. The black tiles on the forward section of the OMS pods were added in response to the localized heating noted in this area during earlier flights.

NASA

"Columbia" nearing Pad 39A in preparation for its fifth flight. By this mission, the white paint on the ETs had been eliminated, thus saving 595 lbs. and $15,000 per tank.

Gerald Balzer collection

"Columbia" departing Launch Pad 39A at the beginning of STS-2. This mission marked the first time a "used" spacecraft had undertaken a second manned mission.

Gerald Balzer collection

"Enterprise" moments before touching down on Edwards AFB's concrete runway during the final Approach and Landing Test (ALT) on October 26, 1977. Noteworthy is the "split"rudder serving as a speedbrake, the pitot boom peculiar to "Enterprise", and the photo-reference markings on the main gear tires.

NASA via Dennis Jenkins

"Enterprise" at Dulles International Airport near Washington, D.C. during arresting gear tests. Vehicle was winched through the net by cable connected to nose gear strut.

NASA

"Enterprise" during barge trip from Huntsville, Alabama to the New Orleans World's Fair. It was displayed at the fair from May 12 through November 11, 1984.

The Propulsion/Avionics package module was to be housed in a lifting-body aeroshell which had a hypersonic L/D of 0.83. Two options were presented for landing, either tail-first vertically onto a tripod gear assembly, or horizontally onto a quadrapod impact attenuation system.

Class I shuttle derived vehicles simply replace the orbiter with a Cargo Element. This vehicle will require minimal modifications to existing facilities, and currently is being promoted by MSFC under the name *Shuttle-C*. Payload capability for *Shuttle-C* to Space Station orbits range from 100,000 lbs. with two SSMEs, to 170,000 lbs. when equipped with three main engines. The Cargo Element would weigh approximately 70,000 lbs., and be capable of carrying cargo 15 ft. in diameter and 82 ft. long. The main engines would be attached to an orbiter thrust structure. Although this is an expensive, robust piece of equipment—built to withstand multiple launches—its use precludes major costs to design and qualify a new, "cheaper" design. Other orbiter equipment, such as APUs, and the RCS, would be built minus the "man-rated" redundancy in order to minimize costs. *Shuttle-C* would use three GPCs, two as a redundant pair, and one running "backup" flight software. The main engines and GPCs would be flight-rated orbiter components that have reached the end of their economical service life, and would not be recovered. It is anticipated that a flight rate of three or four *Shuttle-C* flights per year could be sustained in this manner. Because of its high commonality with the existing Space Transportation System, *Shuttle-C* could be operational within three to four years of Authority to Proceed. An engineering development model, using the MPTA-098 thrust structure, is currently being asembled by MSFC.

Class II configurations replace the SRBs with liquid rocket boosters using four space shuttle main engines. These boosters would give the orbiter the capability to carry 100,000 lbs. into orbit, although it still would be constrained to the 15 by 60 ft. payload bay. Other proposals call for using fuels besides LH_2 in the LRBs, such as methane or propane, or a combination (hybrid) using LH_2 and either RP-1 or JP-7. Because of the more delicate nature of the liquid motors, clamshell closures over the aft skirt opening and retrorockets were baselined in addition to parachutes. A flyback capability for the boosters, using Pratt & Whitney JT9D-7R4D turbofan engines and fixed straight-wings also was studied. The LRBs would require extensive modifications to shuttle launch facilities.

Class III studies combined the improvements of Class I and Class II, and resulted in a payload of up to 250,000 lbs. This configuration would be the fastest and easiest way for the U.S. to recover the heavy lift capability abandoned with the *Saturn V*, but is unlikely to win approval.

The Class IV configuration simply removes the main engines, and their associated avionics from the orbiter, and installs them in a recoverable module under the external tank. This frees up some additional room on the existing orbiters, but offers no real performance improvement. A slight variation on this theme is to leave engines on the orbiter, and install additional engines under the ET to increase payload capability. Either of these options will require substantial modifications to the launch facilities, and they are unlikely to be approved.

Another proposal involved the addition of an 27.9 ft. extension to the aft (bottom) end of the ET for use as an Aft Cargo Carrier (ACC). This would double the available payload volume, and accommodate cargo up to 25 ft. in diameter, compared to 15 ft. in the payload bay. The ACC would add 3,300 lbs. to the empty weight of the shuttle system. The most favored use of the ACC appears to be carriage of an Orbital Transfer Vehicle (OTV) to be extended by the Remote Manipulator System (RMS) for use by the space station.

SHUTTLE CARRIER AIRCRAFT

When the air-breathing engines finally were deleted from the orbiter design during 1974, NASA was faced with the question of how to conduct the atmospheric test flight series, as well as how to ferry the orbiter from a remote landing field back to the launch site. John Conroy's (co-developer of the *Guppy* series aircraft) answer was to modify one of the "jumbo-jet" type aircraft into a "Shuttle Carrier Aircraft" (SCA). Proposals were received from both jumbo-jet manufacturers, with Lockheed initially proposing a twin-fuselage variation of the C-5 military transport that carried the orbiter suspended between the fuselages, and Boeing wanting to carry the orbiter on top of a 747. Although the drop-tests would be easier from the Lockheed aircraft, it presented several significant problems: it was too wide for any known runway, and it was very expensive. Lockheed later submitted a proposal to carry the orbiter on top of a C-5A. This second Lockheed proposal, and the one from Boeing were deemed roughly equal in the ensuing technical evaluation, and as late as mid-March 1974 no final decision had been made. The availability of low-cost used 747 aircraft, and the relative scarcity of C-5As, finally drove the selection of the Boeing aircraft. At the time, thought also was given to carrying the external tanks to Vandenberg on top of the SCA. Wind tunnel tests later indicated the drag of the ET would pose a serious safety hazard to the SCA, and the tanks were shipped by barge through the Panama Canal.

On July 18, 1974 NASA purchased Boeing 747-123 N-9668, from American Airlines. This was the 86th 747, having been delivered to American on October 29, 1970. At the time of NASA's acquisition it had logged 8,999 hours during 2,985 flights, mainly between New York and Los Angeles.

Under a $30 million contract from Rockwell International, a Boeing Aerospace/Boeing Commercial Airplane team began modifications to the aircraft on August 2, 1976. The orbiter would be carried on top of the 747, mildly reminiscent of the two-stage Phase A/B concepts. The orbiter would attach to struts at three points, one forward and two aft, that matched the socket fittings built into the orbiter for attachment to the External Tank.

The modifications peformed by Boeing were of two types: Type-one, which are permanent structural and systems changes; and Type-two which are removable structures used only during transporter missions.

Type-one changes included:

- Installation of bulkheads to strengthen the aircraft body.
- Placement of skin reinforcements at critical stress areas.
- Reinforcement of the horizontal stabilizer structure.
- Installation of fittings to attach the struts which hold the orbiter.
- Installation of L-band telemetry equipment and a C-band transponder.
- Installation of a 747-200 series rudder ratio changer.

Type-two modifications included:

- A telescopic orbiter forward support assembly, to be used during the Approach and Landing Tests. This was a bipod consisting of two tubes; each 13 ft. long, and associated end fittings; it had an adjustable drag

Lockheed-California Co.

"Challenger", as STA-099 (Static Test Article) at the Lockheed structural test facility at Palmdale, California. Test rig weighed 43 tons. It later would be rebuilt into OV-099.

Rockwell International

"Challenger" roll-out at Palmdale following conversion from STA-099 to OV-099. Noteworthy is special multi-wheel Shuttle trailer.

Dennis Jenkins

"Challenger" being moved from Palmdale to Edwards AFB. Route was through the streets of Palmdale and Lancaster, then across a special dirt/gravel road to Edwards.

NASA

Only the shuttles "Enterprise", "Columbia", and "Challenger" (shown) were moved by truck overland the 36 miles from Palmdale to Edwards AFB.

Erik Simonsen collection

November 30, 1982 "Challenger" roll-out from VAB on its way to Pad 39A at KSC for STS-6.

strut on the aft side of each tube to support the bipod after orbiter release.

- A fixed orbiter forward support assembly for use during ferry missions. This also was a bipod, consisting of two tubes 8.5 ft. long and associated end fittings; it did not use drag struts.
- Aft support assemblies, each consisting of a drag strut 12 ft. long and a vertical strut 4.5 ft. long, and their associated fittings. The right aft support was fitted with a non-adjustable 4.8 ft. side strut. The left aft support was fitted with dual adjustable side snubbers which were non-load bearing.
- Horizontal stabilizer tip-fins, each measuring 10 by 20 ft. These fixed-position fins gave the aircraft added aerodynamic stability when it was carrying the orbiter. Each fin had a drag strut connected to the upper surface of the horizontal stabilizer. In practice these fins never were removed from the aircraft.

The aircraft's longitudinal trim system was modified to permit two degrees more trim in order to counteract a nose-up tendency caused by the downwash off the orbiter's wing onto the 747's horizontal stabilizer. Additionally, most of the interior of the aircraft's lower (main) deck was stripped, though a number of seats were retained to transport support personnel on ferry missions. In all, the modifications added 11,500 lbs. to the 747 empty weight. Under a separate contract, the 747's Pratt & Whitney JT9D-3A engines were converted to a JT9D-7AHW configuration, increasing take-off thrust from 43,500 lbs. to 46,950 lbs.

The Separation Monitoring and Control System (SMCS) used for the ALT program was installed on the main deck of the aircraft and separation displays and controls were added at the pilot stations. An orbiter-to-ground S-band relay link and a SCA-to-orbiter intercom also were carried during the ALT program.

The orbiter's mated location on the SCA was selected based on considerations of static stability and control, structural modifications required, weight, and mission performance. Center of gravity limits for the SCA with the orbiter mated are 15% of the SCA mean aerodynamic chord (MAC) for the forward limit, and 33% MAC for the aft limit. Longitudinal stability ballast is carried by the SCA to ensure center of gravity limits are not exceeded. The ballast is carried in standard 747 cargo containers in the forward cargo compartment.

On January 14, 1977 Boeing turned the aircraft back to Rockwell for acceptance testing. Upon completion of weight and balance checks, and a limited flight test series, the aircraft was delivered to NASA with the new registration N-905NA. After its return to NASA, a crew escape system was deemed desirable, and the conceptual design for such a system was accomplished by Teledyne McCormick Selph under contract to the Johnson Space Center, which owns N-905NA. The system is designed to allow crew members to safely abandon the aircraft in the event of an emergency. It consists of a sixteen foot long escape slide connecting the flight deck with an egress port located in the forward cargo bay. In the event of an emergency, a crew member activates the sytem by pulling an initiation handle, one of which is located on either side of the autopilot control pedestal. Upon activation, 30 fuselage windows are explosively fractured to allow immediate aircraft decompression. Three seconds later the emergency egress port is explosively severed and blown clear of the fuselage and a spoiler is extended into the airstream to allow crew members to sufficiently clear the engine nacelles, landing gear and empennage. Crew members then descend using individual parachutes.

Boeing also designed and built the orbiter "boat-tail" tailcone fairing used during early approach and landing tests, and on ferry flights. This fairing significantly reduces aerodynamic drag by smoothing the orbiter's SSME and OMS engines. The tailcone is 36 ft. long, 25 ft. wide and 22 ft. high. It is constructed primarily of aluminum and weighs 5,768 lbs. Eight adjustable steel fittings connect to attach points on the orbiter. Total cost for development and production of the fairing was $4.2 million.

During February 1988 NASA announced plans to acquire a second used 747 to serve as a backup to N-905NA as a shuttle carrier aircraft. This was, in part, due to the realization that a substantial number of flights will continue to land at Edwards AFB, and partially in response to a recommendation made by the Roger's Commission after the *Challenger* accident. The aircraft selected was a 747-100 that Boeing Military Airplane Company purchased for NASA from Japan Air Lines (JAL). Modifications similar to those made to N-905NA are expected to be made to the new aircraft by Boeing, and a critical design review is scheduled to be held during January 1989. The aircraft currently is in storage at San Antonio, Texas, and is not expected to enter operational service until early 1992.

ENTERPRISE

On June 4, 1974 workers at the Air Force Plant #42, Site 1, in Palmdale, California, started structural assembly of the crew module of the first space shuttle orbiter. This vehicle was designated OV-101 (Orbiter Vehicle-101) by NASA, and was to carry the name *Constitution* in honor of the U.S. Constitution's Bicentennial. However, some groups had other ideas about the name for the first reusable manned spacecraft; fans of the TV science-fiction show *Star Trek* staged a write-in campaign urging the White House to rename OV-101. The "Trekkies" had their way, and when the doors opened on September 17, 1976, the name on the side of the orbiter was *Enterprise*.

OV-101's first tasks would not carry her into space, therefore, she did not carry much of the equipment and systems required for space travel. *Enterprise* contained no Main Propulsion System (MPS) plumbing, or internal fuel lines and tankage. The Space Shuttle Main Engines (SSME) and their expansive nozzles on the tail were mock-ups, as were the Orbital Maneuvering System (OMS) pods beside the vertical tail. Ballast was used to maintain the vehicle's weight and balance within the limits anticipated for space-rated orbiters. *Enterprise's* fuel cells were fed hydrogen and oxygen from high pressure tanks instead of cryogenic dewars. The payload bay contained no structural supports or mounts for payloads, and the payload bay doors did not contain any hydraulics or radiators. The crew module contained only a limited flight deck, without the mid-deck lockers, galley, shower and other crew support items to be installed for space flight.

NASA via Gerald Balzer

The decision to delete a proposed air breathing engine option from the orbiter configuration meant NASA had to find another method of conducting the initial flight test series and means of inter-site transport. Although both the Lockheed C-5A and Boeing 747 were examined, the scarcity of the Lockheed aircraft dictated the 747's selection.

Gerald Balzer collection

"Enterprise" sporting the special empennage fairing developed specifically for orbiter ferry flights between sites. The fairing lowers empennage section drag and facilitates improved Boeing 747 directional stability. Noteworthy are large tip plates attached to the 747's horizontal tail surfaces to increase vertical surface area.

NASA

The newest orbiter, OV-105, under construction at Rockwell International's Palmdale, California facility during late 1988. This orbiter will replace "Challenger" when it enters the inventory during early 1990. Green color is the result of preservative covering being applied to natural metal materials.

Rockwell International

"Columbia" cockpit roof area. Markings are applied to each of the Thermal Protection System tiles. Each tile is shaped to fit a unique spot on the orbiter. Weighing only 9 lbs. per cubic ft., the high-temperature black tiles can withstand 2,300° F. The white tiles are used in lower temperature areas and can withstand up to 1,200° F.

Dennis Jenkins

"Challenger" being rolled back to the VAB from Pad 39A after the first attempt to launch STS-26.

NASA via Erik Simonsen

"Challenger" during initial ascent from the KSC on STS-6, its first flight.

NASA via Gerald Balzer

"Challenger" during its second flight (STS-7). RMS arm is visible left of bay. Photo taken from SPA-01.

NASA via Gerald Balzer

"Challenger", on April 9, 1983, shortly after touching down at Edwards AFB following STS-6, its first flight. Roll-out distance was 7,300 ft.

NASA via Gerald Balzer

"Challenger" returning from its last successful flight, STS-30, on November 6, 1985. A record-setting eight crew members were aboard.

NASA via Gerald Balzer

Proof-of-concept night landing was undertaken by "Challenger" at the end of STS-8 on September 5, 1983 at Edwards AFB on Runway 22.

NASA via Gerald Balzer

"Challenger" at the Shuttle Landing Facility at KSC after the STS-11 flight. This marked the first time a shuttle had returned to KSC instead of Edwards AFB.

NASA via Erik Simonsen

"Challenger" arriving at the Kennedy Space Center on July 5, 1982. It was moved immediately into the demating complex and off-loaded for movement to the VAB.

Dennis Jenkins

"Challenger" being uploaded into the mate/demate facility at Edwards AFB, prior to its first flight to the Kennedy Space Center.

The Thermal Protection System (TPS) tiles were simulated by white and black polyurethane foam bonded to the vehicle's aluminum skin.

On the flight deck, the pilots faced an instrument panel decidedly less complex than the later orbiters. Most of the navigation, guidance and propulsion instrumentation was missing, along with the star tracker. Three cameras recorded the pilot's action and Lockheed zero-zero ejection seats were provided in the event escape was necessary. Two blow-out panels were installed above the pilots to facilitate ejection, or rapid escape on the ground. The two windows looking from the aft flight deck into the payload bay were absent, covered by aluminum panels, as were the overhead rendezvous windows. The landing gear system was lowered by explosive bolts and gravity, and there were no provisions for manual or hydraulic retraction.

While Rockwell was busy building OV-101, NASA was defining a flight test series to verify the subsonic airworthiness of the orbiter. As initially envisioned, these tests would take four forms:

- **Taxi Tests.** These would be used to verify the very low-speed dynamics of the mated 747/orbiter combination. The taxi tests would use the concrete runway (04/22) at Edwards, and be conducted in the early morning to minimize problems associated with heat build-up in the SCA's tires and brakes.
  The first run would take the mated vehicles to 75 knots, then be stopped using normal braking. The second test would get up to 120 knots and also stop using normal braking. The last run would be at 135 knots and use full braking, thrust reversers and speed brakes to simulate an aborted take-off.
- **Captive-Inert Flights.** Six flights with an unmanned inert orbiter were planned. These tests would verify the performance, stability and control, flutter margin and buffet characteristics of the mated configuration in flight patterns similar to the manned free flights. The orbiter would weigh 150,000 lbs. resulting in a combined weight of from 585,000 to 630,000 lbs. depending on the 747 fuel load. Flights would be conducted up to 25,000 ft. and 275 knots.
- **Captive-Active Flights.** Astronaut crews would be aboard the orbiter during the six active-captive flights which were designed to determine the optimum separation profile based on the captive-inert test results, refine and finalize orbiter and SCA crew procedures, and evaluate orbiter integrated systems operations. Five of the flights were scheduled with the orbiter tailcone attached, and one without. The mated combination would fly a "race-track" course 40 miles by 14 miles in altitudes up to 24,000 ft. Speeds would range from 235 knots to 260 knots.
- **Free-Flights.** A series of up to eight free-flights were scheduled to follow the first five captive-active flights. The free-flights were designed to verify orbiter subsonic airworthiness, integrated system operations, and both pilot guided and automated landing capabilities. The first five free-flights would be conducted with the tailcone on. Following the successful completion of these five flights, the last captive-active flight (tailcone-off) would be conducted. If this flight was successful, the last three free-flights with the tailcone off would be flown.
  The tailcone-on flights would involved the separation of the orbiter from the SCA at 250 knots at 22,000 ft. (AGL). The orbiter would glide for about 25 miles on its way to a landing, which would be either on the lakebead or the concrete runway.
  The tailcone-off flights would be conducted at 18,000 ft. and due to the increased drag resulting from the lack of the tailcone, the orbiter would only glide about 12 miles. Landings would all be made on the lakebed.

On January 31, 1977 *Enterprise* was towed 36 miles through the California desert from her Palmdale birthplace to Edwards AFB. The trip took the better part of the day, using mostly public roads that had been cleared of traffic, and a few specially constructed gravel roads to provide a shortcut around one of Edward's dry lakes.

Enterprise was mated to N-905NA on February 8, 1977 using the mate/de-mate facility installed at the Dryden Flight Research Center. The mated pair completed weight, balance, vibration and three taxi-tests on February 15, 1977. On February 18, 1977 the orbiter and its carrier aircraft lifted off and were airborne for the first time, reaching an altitude of 28,565 ft. The pilot, Fitzhugh Fulton, Jr. and copilot, Thomas McMurtry, of the SCA were pleasantly surprised to find the orbiter had little adverse effect on the handling of the 747. Several flight profiles were flown, including engine-out simulations, and the last of these flights lasted 1 hr. and 39 min. on March 2, 1977. None of the five captive-inert flights revealed any stability or flight control problems.

On June 18, 1977 veteran astronaut (*Apollo 13*) Fred W. Haise, and test pilot Charles Gordon Fullerton were aboard *Enterprise* when it and the SCA lifted off from Edwards at just past 0800 hours. This test was the first

NASA via Gerald Balzer

Approximately 59.82 seconds after launch, a 70mm still photograph of "Challenger" during STS-33 taken from Camera Pad 10 shows burn-through plume on right SRB.

in which the orbiter was powered up, and Haise and Fullerton practiced moving the orbiter's aerosurfaces and split rudder/speed-brake for the first time. During a flight lasting 55 minutes and 46 seconds, no problems were uncovered. The next flight, with Joe H. Engle and Richard H. Truly at the orbiter's controls, was designed to simulate the separation maneuver that was to be used for the first drop test. This flight took place on June 28, 1977, and the SCA crew pushed the mated ships through climb, pitch down and acceleration tests between 22,000 and 20,030 ft. During this flight, the *Enterprise* crew powered on the autoland system and microwave scan-beam landing system and flew through a simulated precision approach. This second flight lasted one hour and two minutes. A third captive-active flight was flown on July 26, 1977 with Haise and Fullerton, lasting 59 minutes and 53 seconds. After landing, while still mated to the SCA, Haise deployed the *Enterprise's* landing gear as a final check prior to the first separation. These flights revealed no reason not to proceed with the free flights, and the last two tailcone-on captive-active flights were cancelled.

During the week of August 8, 1977, project officials concluded a two-day shuttle readiness review. All conditions were "go". Shortly after 0800, August 12, Fitzhugh Fulton, Jr. and Thomas McMurtry guided N-905NA down Runway 22 with Haise and Fullerton back at the controls of *Enterprise*. At 0848, while flying at 270 knots, Fulton pushed the nose of the 747 down seven degrees, and Haise detonated the explosive bolts holding the pair together. *Enterprise* separated cleanly and Haise put the orbiter into a right-hand turn and pitched up. Using the side-arm controller, Haise held two degrees of pitch for three seconds, then banked 20° to the right and towards Runway 17. The orbiter maintained a 9° nose down attitude as Haise and Fullerton executed two 90° turns. On final, Haise pointed the orbiter down the centerline and opened the split-rudder speedbrake in the vertical tail. Descending at one foot per second, the rear tires hit the lakebed at 185 knots, and the orbiter rolled 11,000 ft. with minimal usage of the wheel brakes. The first flight of *Enterprise* had lasted just over five minutes, and no serious problems were discovered.

Engle and Truly were aboard when *Enterprise* flew again on September 13, 1977 for a series of more extensive maneuvers. *Enterprise* was stable as it began a wide 55° turn with its nose angled up three degrees. Truly took control and flew the vehicle down to 2,000 ft., handing off to Engle for a final series of maneuvers with the speed brake open 40%. Five minutes and 28 seconds after leaving Fulton and McMurtry in the 747, the orbiter's main gear settled onto the lakebed, rolling out 10,037 ft. in just over a minute. The most serious problem to come along during the flight was not in either of the vehicles: the radar at Dryden failed 28 minutes into the captive portion of the flight, and almost caused an abort before it was brought back on-line. The astronauts reported that the use of elevons after nosewheel touchdown was not effective in steering the orbiter, but other than that, the vehicle continued to perform well in flight.

On the third flight, conducted September 23, Haise and Fullerton tested the autoland system, letting the GPCs fly the orbiter down to the 900 ft. pre-flare. The crew hard-braked the vehicle on Edward's lakebed runway 17 without any problems.

The first three free flights were deemed so successful that the final tailcone-on flight was cancelled. The original plan of conducting a mated flight with the tailcone off also was scrubbed. If the buffeting from the mated vehicles became excessive, they could simply abort the flight and land on the lakebed. The time had come to see how *Enterprise* would fly in a return-from-space configuration.

Free-flight number four took place on October 12, and this time the orbiter flew like the brick it was, with a free-flight time nearly half of what it was with the tailcone on. The buffeting in the 747 as the orbiter pulled away was moderate but acceptable, and the handling qualities of the orbiter were basically the same, with the exception of the more rapid descent. Engle and Truly experienced some problems with the TACAN system, and some minor error messages from the GPCs, but the flight cleared the way for the final test: landing *Enterprise* on Edward's concrete runway.

The fifth, and last, flight from the SCA began on October 26, 1977 with Haise and Fullerton at the controls of *Enterprise*. The orbiter separated 51 minutes after take-off at an altitude of 19,000 ft. As before, the crew put the orbiter through a series of maneuvers, finding again that the gliding performance of the orbiter was better than predicted. This time, however, as the crew set up their final approach, trouble started. Coming out of the pre-flare, *Enterprise* was dropping at 290 knots, considerably quicker than planned and, in an attempt to slow the orbiter, Haise opened the speed brake early. Instead of slowing down, the speed increased, so Haise deployed the landing gear and pitched the nose down to make the runway impact point. Since *Enterprise* was unpowered, there was no option of making a second pass. The orbiter's wings dipped and Haise struggled to right the plane as the rear wheels hit the tarmac hard. Instead of dropping her nose, *Enterprise* suddenly took off again, bouncing 20 ft. into the air. After a few anxious seconds, the orbiter smoothed out and stabilized for the remainder of the roll-out, with the nose finally dropping onto the runway.

Review of flight data indicated that earlier use of the speedbrake would have reduced the airspeed problem. Also, Haise had lowered the elevons, which caused an increase in lift as the ballooning began after touchdown.

NASA via Gerald Balzer

The recovered remains of "Challenger" following the failed STS-33. Approximately 30% was recovered.

NASA via Gerald Balzer

The "Challenger's" main engines. Because of their easy detectability, most SSME parts were recovered.

NASA

The 15 in. x 28 in. burn-through point on the right SRB. It was recovered from a depth of 560 ft.

NASA announced that the ALT program had met its objectives, and no further flights would be conducted. The orbiter was aerodynamically cleared for flight into space.

The fifth shuttle free flight concluded *Enterprise's* flight testing. Technicians reinstalled the tailcone and Fulton and the 747 crew completed a series of test flights with the orbiter in a ferry configuration during mid-November 1977. During March 1978, the SCA carried *Enterprise* to the Marshal Space Flight Center in Alabama for the Mated Vertical Ground Vibration Tests (MVGVT). These tests used a set of exciters and sensors placed on the skin of the mated components to create and monitor vibrations and resonances similar to those that later would be encountered in flight.

This sensor system, called the Shuttle Modal Test and Analysis System (SMTAS), provided automatic control of up to 24 preselected exciter channels from the available 36 150 lb. and 20 1,000 lb. exciters. The exciters apply simultaneous precise pressure at preselected points on the vehicle. The mostly enclosed test facility was constructed during 1964 to test the *Apollo/Saturn V* moon-rocket. It was modified during 1975 to 1977, enlarging it from 98x98x360 ft. to 98x122x360 ft. to accommodate the winged space shuttle instead of the cylindrical *Saturn V*. Cost of these modifications were $2,880,000.

The first test configuration included *Enterprise* and an external tank to simulate the high altitude portion of flight after SRB separation. During these tests, the LO_2 tank contained between 3,450 and 101,000 gals. of deionized water, and the LH_2 tank was pressurized but empty. The combined orbiter/ET weighed in at 1,200,000 lbs. and was suspended by a combination of air bags and cables attached to the top of the test stand.

The second test configuration added a set of solid rocket boosters containing inert propellant to simulate lift-off conditions. This marked the first time that a set of dimensionally correct elements of the Space Transportation System had been assembled together. One-hundred-forty-thousand-six hundred gals. of deionized water filled the LO_2 tank. The test article weighed 4,000,000 lbs. during these tests and was supported by four hydrodynamic stands, two under each SRB. Under test conditions these stands contained 1,000 gals. of special oil pressurized to 1,500 psi. Bearings on top of the stands created the "floating" characteristics desired for the tests. These stands originally were used for the *Saturn V* tests, and were refurbished by the Denver Division of Martin Marietta Corporation. This series of tests started on September 20, 1978.

The final test configuration was similar to the second except that the SRBs were empty and the LO_2 tank held only 101,000 gals. These tests simulated the period of flight just prior to SRB separation, and the vehicle was supported using the same system as in the second test series. These final tests were completed on February 26, 1979, and cleared the space shuttle configuration for space flight.

NASA

"Challenger" nose section being recovered on January 30, 1986, by a combined USN and USCG effort.

At this point, *Enterprise* was supposed to have gone back to Rockwell in Palmdale to be refitted as a flight orbiter. Several things convinced NASA to change its mind. As OV-101 was being built, numerous lessons were learned regarding the orbiter's design and the materials used in their construction. Subsequent orbiters (from STA/OV-099 on) would use wings and a mid-fuselage significantly stronger than that installed on *Enterprise*. Some aluminum castings in other areas of the fuselage were changed to titanium as a weight saving measure. The need to retrofit these changes on *Enterprise*, as well as the time and expense required to disassemble her, would lead to a decision to modify the structural test article (STA-099) instead. STA-099 never was completely assembled, so there would not be as much disassembly required, and most of the structural changes had been incorporated during production.

So instead of heading to California, *Enterprise* was loaded aboard the SCA for the one hour and 52 minute flight to the Kennedy Space Center on April 10, 1979. While at KSC she would be used to check out facilities and procedures that would be used to support *Columbia* during her first flight. Prior to *Enterprise's* arrival, a 75 ton "orbiter simulator" had been used to practice lifting and handling the orbiter. This simulator was a steel structure roughly resembling the shape of an orbiter, and was more or less dimensionally correct. Dubbed *Pathfinder*, it was constructed at MSFC in 1977 as a stand in for the *Enterprise*. *Pathfinder* was used to fit-check the roads and facilities at Marshall that were used during the Mated Vertical Ground Vibration Tests, as well as the hoisting system that was used to lift *Enterprise* from the back of the SCA. (Unlike the Mate/Demate devices at KSC and Dryden, the MSFC hoist is transportable, and was used during the contingency landing of STS-3 at White Sands). During mid-1978, *Pathfinder* was shipped by barge to KSC where it was used to fit-check the Mate/Demate device, the OPF and the VAB work platforms, and for ground crew training. During early 1979, *Pathfinder* was used to rehearse post-landing procedures at the KSC Shuttle Landing Facility. Later during 1979 *Pathfinder* was returned to Marshall. After several years in storage, the American-Japan Society expressed an interest in obtaining a full size orbiter model for display during the "Great Space Shuttle Exposition" to be held in four Japanese cities, including Tokyo and Osaka, from June 1983 to August 1984. The Japanese agreed to provide approximately $1,000,000 to have Teledyne-Brown Engineering modify the ungainly orbiter simulator to more closely resemble an orbiter. Following display in Japan, *Pathfinder* was returned to MSFC for permanent display at the Space and Rocket Center near Huntsville, where it is mounted on a special platform, mated to the external tank used during the main propulsion tests (MPTA-ET), and a pair of solid rocket boosters. These SRBs are assembled from composite casings produced during the abortive attempt to build filament-wound motor casings for Vandenberg, as well as various engineering and structural test pieces, such as nose segments and aft skirts.

NASA via Gerald Balzer

A 9 ft. 7 in. x 16 ft. segment of "Challenger's" right wing following recovery from some 70 ft. of water.

OV-101 was mated to a pair of inert solid rocket boosters and an external tank and transported to launch pad 39A on May 1, 1979. During almost three months at the pad, *Enterprise* would help verify that maintenance platforms mated to the vehicle in the correct locations, and that crew escape procedures worked properly. On July 23 OV-101 was rolled back to the VAB to be demated from the SRBs and ET. During August the vehicle was flown to Vandenberg AFB to let the workers building the launch complex there have a look. On August 16, 1979 OV-101 was flown to Edwards where it was demated from the SCA, and on October 30, moved overland back to Palmdale. While at Palmdale, select OV-101 parts were removed and refurbished for use on later orbiters as a cost saving measure. Wearing a new coat of paint, it was returned to Edwards on September 6, 1981.

During the May-June 1983 time period, *Enterprise* was ferried to France for an appearance at the Paris air show. During the trip the vehicle also visited Germany, Italy, England and Canada. It marks the only time to date that an orbiter has been outside the United States. During the trip, the Shuttle Carrier Aircraft was fitted with infrared and radar warning devices and countermeasures to guard against terrorist attack.

On April 5, 1984, *Enterprise* arrived via barge at the World's Fair in New Orleans. A special stand had been built, and cranes carefully lifted the orbiter onto it. OV-101 would remain on display through the end of the Fair.

During late 1984 and early 1985, *Enterprise* was again called to perform in its role as pathfinder. Mated to a pair of inert solid rocket boosters, and the external tank scheduled for STS-3V, *Enterprise* performed in a series of tests known as FVV (Flight Vehicle Verification) at the Vandenberg Launch Site. These tests verified that all facilities at Space Launch Complex Six (SLC-6) and in the Orbiter Maintenance and Checkout Facility (OMCF) would support the planned launch of OV-103 during 1986. During early 1986, *Columbia* was scheduled to go to the Vandenberg Launch Site to conduct tanking tests and a flight readiness firing to complete validation of the launch complex. This was cancelled as being "too risky" and too expensive (at $60-million) after the loss of *Challenger*. *Discovery* was to fly the first mission from Vandenberg and be permanently based there when the site became operational.

After completion of the Vandenberg tests, *Enterprise* was ferried on May 24, 1985 from Vandenberg to Dryden. Following a ferry flight on September 20, 1985, OV-101 was placed on display for several months at the Kennedy Space Center next to a *Saturn V* moon rocket while a storage place at Dulles Airport in Washington, D.C. was prepared. After transfer to the National Air and Space Museum on November 18, 1985, *Enterprise* was tasked

"Columbia" being moved overland from Palmdale to Edwards AFB. The move took most of one day to complete.

Rockwell International via Gerald Balzer

"Columbia" mounted on the back of NASA's Boeing 747-123, N-905NA, shortly before its initial delivery flight to the Kennedy Space Center.

Rockwell International via Gerald Balzer

to perform one more test for NASA during the week of June 8, 1987. A landing arresting barrier, similar to ones used by the AF and Navy to catch damaged aircraft, was erected at Dulles, and *Enterprise* was slowly winched into it to determine if an orbiter could successfully use one. Although the tests were deemed successful, there are no plans to erect such barriers at either of the primary landing sites (Edwards and KSC).

During October 1986, *Atlantis* was used as a verification vehicle to validate the second launch pad at KSC (Pad 39B) in preparation for its first use during Flight 25. In addition to the normal lift checks and interface verifications, a new weather protection system successfully was tested, and this subsequently was retrofitted to Pad 39A.

OTHER TESTING

Enterprise was not the only test article built. At least two other partial airframes were constructed; MPTA-098 was a Main Propulsion Test Article, and STA-099 was a structural test airframe that later was modified into OV-099. An upper forward-fuselage half also was constructed and sent to Holloman AFB on October 26, 1976 for rocket-sled testing of the shuttle ejection seats. Testing commenced on November 18, 1976 and several unmanned ejections were accomplished at various speeds, clearing the use of the ejection seats during the ALT and orbital flight tests.

The main propulsion test article consisted of an orbiter aft-fuselage, a truss arrangement which simulated the orbiter mid-fuselage, and the SSME thrust structure including all main propulsion system plumbing and electrical systems. MPTA-098 was completed and shipped to the National Space Transportation Laboratory (NSTL) on June 24, 1977. NSTL now is known as the Stennis Space Center. There it was mated with an external tank (MPTA-ET) and three SSMEs for propellant loading and engine firing tests. The first static engine firing took place on April 21, 1978 and lasted for 2.5 seconds. By July 7 the firings were up to several minutes in length, and included engine restart demonstrations. These early firings were with non-flight rated engines. By the summer of the next year, flight qualified hardware had been installed on the MPTA, and testing continued but not without its problems. On July 2, 1979 a main fuel line ruptured nineteen seconds into a test. During October, an over-sensitive hydrogen leak sensor terminated a firing before it began. The next month a LO_2 high-speed turbopump failed after ten seconds. On December 17 a successful static firing took place involving three non-flight rated engines cycling between 70% and 100% thrust for 554 seconds. Two more successful firings would be conducted during early 1980. Then on April 16, an overtemp of the hydrogen high-speed turbopump caused the number 2 engine to shut down 4.6 seconds into a 544 second test. Another success was followed by another failure. The tenth static firing, this time with flight nozzles, was shutdown 105 seconds into the test due to a burn-through in the engine number three hydrogen preburner. The SSMEs were becoming the pacing item towards OV-102's first flight. On January 17, 1981, just three months prior to the scheduled launch date, MPTA-098 successfully demonstrated a 625 second flight, with simulated abort profiles. The flight-rated engines were flown at 100% thrust, and included gimbal checks of the thrust vector control system. This test also included loading an ET without the LO_2 anti-geyser line, to verify the feasibility of removing it on the "lightweight" ETs.

While *Enterprise* and MPTA-098 were undergoing tests, Lockheed in Palmdale was busy trying to verify the structural integrity of the Structural Test Article (STA-099). On February 4, 1978, Rockwell had delivered a complete airframe named *Challenger* to the Lockheed-California Company located across the runway in Palmdale. Twelve months of stress and structural testing would take place in a 43 ton steel rig built especially for the shuttle test program. The test rig contained several hundred hydraulic jacks which simulated various stress levels under control of a computer. These stress levels duplicated the launch, ascent, orbit, reentry and landing phases of flight. Three 1,000,000 lb. force hydraulic cylinders were used to simulate the thrust from the boost engines. Heating and cooling tests were conducted along with the stress tests; gaseous nitrogen was used to simulate the cold of space flight, while heating blankets developed for the A-12/SR-71 test program were used to simulate ascent and reentry heating. Twenty-five different heating and cooling zones were established along the length of the fuselage.

Testing was completed successfully and STA-099 was returned to Rockwell on November 7, 1979 for conversion into OV-099. This conversion, while easier than it would have been to convert *Enterprise*, still involved a major disassembly of the vehicle. *Challenger* had been built with a simulated crew module, and the forward fuselage halves had to be separated to gain access to this. Additionally, the wings had to be modified to incorporate lessons learned from the static testing just completed. *Challenger* would end up some 2,000 lbs. lighter than *Columbia*, in spite of having additional operational equipment installed, including two heads-up displays in the cockpit (these would become standard on all orbiters).

Rockwell's original $2.6 billion contract had authorized the building of two static-test articles (MPTA-098 and STA-099) and two flight-test vehicles (OV-101 and OV-102). The 1978 decision not to modify *Enterprise* (OV-101) from its ALT configuration left only one space-rated orbiter. The final assembly of this vehicle, *Columbia*, started on November 7, 1977, and led to a roll-out at the Palmdale plant on March 8, 1979. Meanwhile, on February 5, 1979 NASA awarded Rockwell a supplemental contract to modify *Challenger* (STA-099) into a space-rated orbiter (OV-099), and to construct two additional orbiters (OV-103 and OV-104). This $1.9 billion contract also covered modifying OV-102 following the orbital flight-test series (STS-1 through STS-4). *Discovery* (OV-103) and *Atlantis* (OV-104) would benefit from lessons learned while constructing and testing the first orbiters, and end up weighing some 7,084 lbs. less than OV-102. *Discovery* was rolled out on October 16, 1983, and *Atlantis* followed on April 6, 1985. The experience gained during the orbiter assembly process enabled OV-104 to be completed with a 49.5% reduction in man-hours compared to OV-102.

When *Columbia* arrived back at Palmdale during mid-1984, the planned 24 modifications became more than 240 modifications and improvements. They included structural strengthening to enable it to withstand higher aerodynamic loading, wiring changes, and avionics updates. During *Columbia's* refit, a special instrumentation package was installed to measure the orbiter's aerodynamic and thermodynamic characteristics. The instrumentation consisted of three experiments developed by the Langley Research Center and were part of the Orbiter Experiments Program (OEX) managed by NASA's Office of Aeronautics and Space Technology (OAST). The first flight of the three experiments was on Flight 24 (61-C) during January 1986.

The Shuttle Entry Air Data System (SEADS), developed by a team headed by Paul M. Siemers III, measured the distribution of air pressure around the orbiter's nosecap during reentry. The SEADS experiment required a new orbiter nosecap that contained 14 penetration assemblies distributed across the nose surface in a cross-shaped pattern. Each penetration assembly contained a small hole through which local surface air pressure was sensed during ascent and reentry. Each orifice was connected to two pressure transducers, one of which was sensitive to high-level pressures (0-20 psia), while the other was sensitive to low-level pressures (0-1 psia). All components

"Columbia" at Palmdale undergoing final assembly in Bldg. 294. OMS pods were mock-ups as the real pods already had been shipped to Kennedy. Shuttle was moved to Edwards shortly after this photo.

NASA via Gerald Balzer

"Columbia" after arriving at Kennedy Space Center for the first time and immediately prior to initial pre-flight preparations.

"Columbia" being prepared for uploading in the VAB at Kennedy Space Center. Hoisting unit was specially developed for this requirement.

of the penetration assemblies and nosecap internal tubing were made of coated columbium superalloy.

The Shuttle Infrared Leeside Temperature Sensing (SILTS) experiment was developed by Langley engineers David A. Throckmorton and E. Vincent Zoby. It was designed to obtain high-spatial-resolution infrared images of the leeside (upper) surfaces of the orbiter's port (left) wing and fuselage during reentry. The information indicated the extent of aerodynamic heating to the leeside surfaces in flight, and could not be adequately simulated in ground facilities. The top section of *Columbia's* vertical tail was replaced with a special pod to house the components of the SILTS experiment. The pod was approximately 20 in. in diameter and was capped at the leading edge with a hemispherical dome that had two infrared-transparent windows. The pod contained an infrared scanning system, a data and control electronics module, and a pressurized nitrogen cooling system. The entire pod, as well as the top ten feet of the tail, was covered with black high-temperature RSI tiles.

The last of the new experiments installed on *Columbia* was the Shuttle Upper Atmospheric Mass Spectrometer (SUMS) developed by Roy J. Duckett and Robert C. Blanchard. SUMS sampled the gases at *Columbia's* lower surface through a small hole located just aft of the nosecap and forward of the nose wheel doors. The instrument measured and identified the quantities of various gases to allow a determination of atmospheric density. The SUMS instrument was a mass spectrometer originally designed for the *Viking Mars* mission.

Two orbiters, *Challenger* and *Discovery*, were modified to enable them to carry the *Centaur* upper stage in the payload bay. These modifications included extra plumbing to load and vent *Centaur's* cryogenic propellants (other IUS/PAM upper stages are propelled by solid motors), and controls on the aft flight deck for loading and monitoring the stage. These modifications cost $5 million per orbiter. *Atlantis* was delivered with a *Centaur* capability.

During the construction of the last two orbiters, NASA had opted to have the contractors manufacture a set of "structural spares" to facilitate the repair of an orbiter if one was damaged during an accident. This structural spares contract was awarded during April 1983 with a value of $389 million. The spares consisted of an aft-fuselage, mid-fuselage, forward fuselage halves, vertical tail and rudder, wings, elevons and a body flap. Following the destruction of *Challenger* during 1986, NASA authorized Rockwell to assemble the spares into another orbiter (OV-105) with a contract awarded on July 31, 1987 valued at $1.3 billion. OV-105 will have all the modifications incorporated into the other orbiters over the past few years built into her. In addition, a few new items will be onboard, including a 39 ft. breaking parachute system, much like that originally deleted during 1984, to further reduce the wear on the braking system. The new orbiter currently is scheduled for its first flight during early 1992.

During March 1988 NASA announced a contest to name OV-105. As a tribute to school teacher Christa McAuliffe, who died on STS-33, the contest is open to all school students, kindergarten through 12th grade, in the United States. In honor of the STS-33 crew, the name *Challenger* has been retired. Entries must have been mailed to NASA by December 31, 1988, and the winner will be announced during May 1989. During May 1988 the Space Station Program Office announced it was considering the purchase of an additional orbiter (OV-106) for use as a space station rescue vehicle. It was felt that "piggy-backing" another orbiter during the construction of OV-105 would be more economical than designing a new vehicle for the rescue function.The expense of procuring another orbiter is still considerable, and it is unlikely the Space Station Program Office could afford this option. During the summer of 1988, the Europeans offered the use of a "Hermes" mini-shuttle as a potential rescue vehicle, and this is the more likely option.

NAMES

NASA decided to continue a naval (and unofficial space faring) tradition and name each orbiter. The theme chosen was to name the orbiters after famous sailing ships. The names are as follows:

OV-099, *Challenger*, after a Navy ship which from 1872 to 1876 made a prolonged exploration of the Atlantic and Pacific oceans. It also was the name of the *Apollo 17* Lunar Module.

OV-101, *Enterprise*, after the starship in the science-fiction television show "Star Trek". Originally was to be named *Constitution* in honor of the U.S. Constitution's Bicentennial. Several naval vessels also have carried the name *Constitution*. A write-in campaign during 1976 produced over 100,000 requests for the White House to override NASA, which it did. *Enterprise* still was appropriate since several naval ships, including the world's first nuclear powered aircraft carrier have carried the name.

OV-102, *Columbia*, after a sailing frigate launched during 1836 that was one of the first Navy ships to circum-

"Columbia" in the VAB. White circles are carrier mount points and larger are propellant disconnects.

"Columbia" in preparation for STS-1 and following first completed Kennedy Space Center VAB assembly.

"Columbia" being rolled-out for STS-1. Trip from VAB to Pad 39A is 3.5 miles (to Pad 39B is 4.25 miles).

"Columbia" in preparation for STS-1. Separation rockets are visible on the nose cone of each SRB.

"Columbia" lifting off during STS-4. This was the final orbital flight test.

With first unpainted ET, "Columbia" clears pad at KSC's Launch Complex 39 during STS-3.

navigate the globe. *Columbia* also was the name of the *Apollo 11* Command Module which carried Neil Armstrong, Michael Collins and Edward (Buzz) Aldrin to the moon for the first lunar landing mission, July 20, 1969.

OV-103, *Discovery*, after two ships, Henry Hudson's which during 1610-1611 attempted to search for a northwest passage between the Atlantic and Pacific oceans and instead discovered Hudson Bay, and Captain Cook's which discovered the Hawaiian Islands and explored southern Alaska and western Canada.

OV-104, *Atlantis*, after a two-masted ketch operated for the Woods Hole Oceanographic Institute from 1930 to 1966 and traveled more than half a million miles during its research trips.

OV-105, will be named during May 1989. The name will continue the famous sailing ship tradition.

THE MISSIONS

Following the missions of the space shuttle can be a frustrating experience. NASA has used at least three different numbering systems durig the program. Prior to the *Challenger* accident there were 25 flights flown, and the popular way to list them is by these sequential flight numbers preceded by "STS" (Space Transportation System), hence Flight 1 originally was known as STS-1. Under this system, the numerically highest mission flown was STS-33; eight flights that were assigned numbers having been cancelled. Flights scheduled from the Vandenberg Launch Site were to start with STS-1V. After the ninth flight, NASA instituted a system where each flight was known by a two number/one letter designation. The first digit indicated the fiscal year of the scheduled launch (4 for FY84); the second digit identified the launch site (1 was the Kennedy Space Center, 2 was the Vandenberg Launch Site); and the letter corresponded to the alphabetical sequence for the fiscal year ("A" being the first mission of the year, "B" being second, etc.). This numbering scheme was dropped after the 1986 *Challenger* accident, the program returning to the original "STS" numbering system. This, however, will also cause some problems since NASA decided that the first flight after the accident would be STS-26 (*Challenger* was the 25th flight launched). To overcome the problem of having two "STS-26" missions (Flight 19/51-F before the accident, one after), the mission is carried internally as STS-26R, the "R" signifying "reflight". The "R" will be added to missions through STS-33R, the number carried by *Challenger* on her ill-fated flight.

Following is a brief summary of the 25 missions launched from April 1981 to January 1986:

Flight 1 — (STS-1) — A Flight Readiness Firing (FRF) was conducted on February 20, 1981. This involved firing the main engines for 20 seconds at 100% thrust while the vehicle remained firmly secured to the launch pad as a final check of all systems. A launch attempt on April 10 was scrubbed due to a timing slew in the orbiter's general purpose computer system. The flight was launched April 12, 1981 as a two-day demonstration of *Columbia's* ability to go into orbit and return safely. Its main payload was a flight instrumentation pallet containing equipment for recording temperatures, pressures and acceleration levels at various points on the vehicle. In addition, there were checkouts of the payload bay doors, attitude control systems and the orbital maneuvering system. This marked the first use of solid rockets on a manned vehicle, and the first time astronauts had piloted a new type of spacecraft on its maiden flight. Post-flight inspection showed *Columbia* had suffered minor damage from an overpressure wave created by the SRBs at ignition, had lost 16 tiles, and had 148 more damaged. Otherwise, OV-102 was in good condition for a "used" spaceship.

Flight 2 — (STS-2) — The original STS-2 launch date of November 4 was delayed by an apparent low reading on a fuel cell oxygen tank pressure transducer. An abort at T-31 seconds on November 11 occurred when clogged fuel filters in the auxiliary power units caused an overtemp condition. *Columbia* was launched successfully the next day after a 2 hour and 40 minute delay to replace a data transmitting unit. It was the first manned flight of a "used" spacecraft. This flight marked the first test of the remote manipulator system, and carried a payload of earth survey instruments, as well as the flight instrumentation pallet. Failure of fuel cell #1 shortened the flight by three days. Modifications to the launch pad's sound suppression water system to absorb the SRB overpressure wave were successful, and little damage was noted when OV-102 returned to Edwards AFB after 36 orbits.

Flight 3 — (STS-3) — The launch of STS-3 was delayed by one hour by a failure of a heater on a nitrogen gas purge line. The longest of the initial test flights, *Columbia* stayed aloft for eight days. Activities included a special test of the manipulator in which the robot arm removed a package of instruments from the payload bay, but did not release it into space. The flight included experiments in materials processing, and thermal testing of the orbiter. The latter was accomplished by exposing the tail, nose and top of the orbiter to the sun for varying periods of time, rolling it in between tests to stabilize temperatures over the entire body. This marked the first flight of an unpainted external tank, saving 595 lbs. Heavy rains at

"Columbia" during final STS-9 preparations at Kennedy Space Center. This mission carried Spacelab aloft, a European Space Agency scientific workshop.

"Columbia" on final approach to Edwards AFB at the end of STS-3. Visible is the split rudder assembly used to help aerodynamically decelerate the shuttle.

NASA via Erik Simonsen

"Columbia" during its return at the end of STS-3. NASA Northrop T-38A, N-961NA, normally assigned to the Johnson Space Center, is seen flying chase.

NASA via Gerald Balzer

"Columbia" following STS-1. Difference between STS-1/2 and STS-3 and subsequent flights is the small OMS pod square black tile area.

NASA via Kennedy Space Center

"Columbia" at the end of landing roll-out at Edwards AFB following STS-4. Barely visible approaching from the right is the truck-mounted egress stair unit.

Rockwell International via Erik Simonsen

Post-flight safing of "Columbia" immediately following STS-1. Umbilicals are connected to unit that clears residual propellants and associated gases from shuttle tanks.

Edwards caused a change in landing sites to Northrup Strip at the White Sands Missile Range in New Mexico. High winds at this site caused the mission to be extended one day. *Columbia* returned home with 36 tiles missing and another 19 damaged.

Flight 4 — (STS-4) — The first space shuttle flight to be launched on schedule, this *Columbia* flight featured another test of the robot arm, which extended a scientific payload over the side of the payload bay, then reberthed it. Materials processing experiments were conducted, as were a number of scientific investigations. This flight carried the first DoD payload. During FY 82-83, the Defense Department had paid NASA a total of $268 million for nine military Space Shuttle launches. The only major problem on this flight was the loss of the two SRBs. The main parachutes failed to function properly and the two casings impacted the water at too high a velocity and sank. They later were found and examined by remote camera, but were not recovered. With the landing of STS-4 on the concrete runway at Edwards, the orbital flight test program came to an end with 95% of its objectives completed. Since more than the two pilot flight crew would be carried on future flights, the ejection seats were disabled after this flight.

Flight 5 — (STS-5) — On its first "operational" flight, *Columbia* launched two communications satellites, which later were boosted into geosynchronous orbit by attached payload assist modules. The crew of four advertised themselves as the "We Deliver" team. This was the second "on-time" launch of the Space Transportation System. A planned Extra-Vehicular Activity (EVA) by Allen and Lenoir was cancelled when a ventilation motor in one suit and a pressure regulator in the other malfunctioned.

Flight 6 — (STS-6) — A Flight Readiness Firing on December 19, 1982 revealed a hydrogen leak in a fuel line to the #1 SSME. A second FRF conducted on January 25, 1983 revealed leaks in all three SSMEs. Repairs were made to the engines and the first flight of *Challenger* was launched on April 4, 1983. It was highlighted by the first shuttle-based EVA. The crew successfully deployed the first Tracking and Data Relay Satellite (TDRS), but after deployment, the upper stage that was to boost the TDRS-A into a higher orbit lost steering control, leaving the satellite in a useless orbit. Nearly two months of short attitude control thrusters bursts eventually nudged it into its correct orbit.

Flight 7 — (STS-7) — This was the third flight to be launched with no delays, and *Challenger* delivered a pair of communications satellites to orbit altitude. Additional tests of the remote manipulator system were performed, releasing and recapturing the SPAS (Shuttle Pallet Satellite) test article. While SPAS was being held by the arm, Bob Crippen fired *Challenger's* reaction control thrusters to observe the effect this had on the extended arm. The first tests using the TDRS satellite deployed on STS-6 were conducted during this flight. This marked the first retrieval of an object (SPAS) from orbit.

Flight 8 — (STS-8) — *Challenger* made the first night launch of the shuttle program, providing a spectacular view for several thousand spectators. The time of launch was dictated by the tracking requirements of the INSAT-1B primary payload. Additional tests of the TDRS system were conducted in preparation for the upcoming Spacelab missions. *Challenger's* nose was held away from the sun for 14 hours to test the flight deck area's performance in the extreme cold. The flight ended in the first night landing, conducted just to make sure it would work.

Flight 9 — (STS-9, 41-A) — The *Columbia*, in its last flight prior to an extended modification period, carried the first Spacelab. Designed in West Germany and funded by the European Space Agency (ESA), Spacelab offered dedicated science research into such areas as materials processing, astronomy, microgravity research, and physics. This marked the first operational use of the TDRS system. Two of the quad-redundant general purpose computers and one inertial measurement unit failed during this flight, causing a 7.5 hour landing delay. Upon landing, several small fires were discovered in the tail section, caused by leaks of hydrazine from the Auxiliary Power Units.

Flight 10 — (STS-11, 41-B) — This *Challenger* mission was highlighted by the introduction of the Manned Maneuvering Unit (MMU), an untethered backpack propulsion unit that allows astronauts to maneuver in space independent of the orbiter. An inflated balloon was to be released from the payload bay as a rendezvous target, but it burst upon inflation. However, it still formed a large enough target, and rendezvous operations in preparation for the Solar Maximum retrieval scheduled for 41-C were completed successfully. The mission also launched two communications satellites, but their boosters failed to ignite, leaving them stranded in low-earth orbit. For the first time, a shuttle landed on the concrete runway at the Kennedy Space Center.

Flight 11 — (STS-13, 41-C) — *Challenger* demonstrated an important capability of shuttle: the retrieval, repair and redeployment of the malfunctioning Solar Maximum Mission spacecraft with the help of the Manned Maneuvering Units. *Challenger* was maneuvered within 200 ft. of the Solar Max, and astronauts Nelson and van Hoften made several attempts to capture the spacecraft unsuccessfully. The next morning, they were successful on the first attempt, capturing and securing the spacecraft in *Challenger's* payload bay for repairs. Solar Max was redeployed the next day, to continue its observations of the sun. This was the first flight to use a "direct ascent trajectory", where the orbiter's main engines carried it all the way to its planned operational altitude of 288 miles. The OMS engines only had to be used once, to circularize the orbit. Other activities included deployment of the Long Duration Exposure Facility (LDEF-1).

Flight 12 — (STS-16, 41-D) — *Discovery* had its FRF on June 2, 1984. The performance was nominal, leading to a planned launch date of June 22 which later was changed to June 25. The launch attempt on that day was scrubbed late in the countdown, due to a failure in the backup GPC. The offending computer was replaced, but another launch attempt the next day was scrubbed by the onboard computers at T-4 seconds when SSME #3 lost redundant control over a main fuel valve immediately after ignition. SSME #2 barely had ignited, and #1 had not, when the shutdown occurred. This was the first on-pad abort-after-ignition of the shuttle program. After this abort, the 41-D mission was remanifested to include the most important payload items from both the originally planned cargo, and that intended for 41-F, which then was cancelled. This required returning the vehicle to the VAB for disassembly and reconfiguration. An August 29 launch attempt failed due to a problem with the computer software for the main engine controllers. *Discovery* finally was launched on her maiden flight on August 30. The mission demonstrated repeated deployment and retraction of a large, foldable solar array to investigate the practicality of using such solar wings as power sources for extended shuttle missions and the space station, and launched three communications satellites.

Flight 13 — (STS-17, 41-G) — *Challenger* launched the NASA Earth Radiation Budget Explorer. A payload bay pallet carried instruments for earth observations, including an advanced imaging radar, and a crew of seven (a record) focused on scientific research. A rehearsal EVA was conducted by Sullivan and Leestma for the refueling of Landsat-4 on a later mission. The crew of seven included two women and a Canadian. The only problem came when the orbiter's Ku-band antenna lost the ability to track, requiring the crew to "point" the orbiter at TDRS instead of the usual procedure of pointing the antenna.

Flight 14 — (STS-19, 51-A) — A one-day slip in the launch date caused by high-altitude "wind-shear" was the only launch problem. *Discovery* deployed two communications satellites, and retrieved two others that had been sent into unusuable orbits after deployment on Flight 10. This was the first time an object placed into orbit by one vehicle had been recovered by another. The retrieval of PALAPA B2 and WESTAR VI involved considerable determination to succeed by astronauts Gardner and Allen during extended EVAs.

Flight 15 — (STS-20, 51-C) — *Discovery* carried the first dedicated DoD payload. The payload was boosted into higher orbit by an attached Inertial Upper Stage (IUS), that, according to official Air Force statements, "... successfully met its mission objectives..."

Flight 16 — (STS-23, 51-D) — This *Discovery* flight was a composite mission, carrying part of its original manifest, and part of that from 51-E, which had been cancelled. The crew was entirely from the cancelled mission, except for Charles Walker, who substituted for Patrick Baudry because the latter's flight experiments no longer were on the manifest. 51-E was to have been a *Challenger* flight, and the crew patch selected for the re-manifested 51-D was that of the cancelled flight with the substitution of *Discovery's* name for *Challenger's* on the patch. The mission also featured the first flight of an elected official, Senator E. J. "Jake" Garn (R-Utah), Chairman of the Senate Committee with oversight responsibilities for NASA's budget. Two communication satellites were deployed, but one, Leasat-3 did not successfully fire its booster stage. *Discovery's* flight was extended for two days to permit the crew to try and work-around the problem using a "fly-swatter" to trigger a malfunctioning switch. Despite repeated efforts, the spacecraft remained lifeless. A tire blew upon landing at the Kennedy Space Center, causing all future landings to be at Edwards until the inactive nose-wheel steering system could be activated and tested.

Flight 17 — (STS-24, 51-B) — The second Spacelab mission and materials processing experiments were carried by *Challenger*. This was the first flight of a fully operational Spacelab configuration. Spacelab capabilities for multi-disciplinary research in microgravity were demonstrated successfully. The gravity gradient attitude of the orbiter proved quite stable, allowing the delicate experiments in materials processing and fluid mechanics to proceed successfully.

Flight 18 — (STS-25, 51-G) — An international flavor was added to *Discovery* with the addition of a Saudi Prince and a French "spatialnaut" to the crew. Three communications satellites, one Saudi, one Mexican, and one belonging to AT&T were deployed. All three reached geosynchronous orbits and entered service with their owners. Spartan-1, a 2,223 lb. free-flying astronomy experiment, also was deployed and retrieved successfully.

Flight 19 — (STS-26, 51-F) — On a July 12 launch attempt, a hydrogen coolant valve in SSME #2 failed to close, causing an on-pad abort at T-3 seconds. Launch was rescheduled for July 29, but was delayed by 1 hour and 37 minutes due to an orbiter problem. After repairs, *Challenger* was launched at 17:00 on July 29, but at T + 5 minutes 45 seconds, the onboard GPCs shut down the Number 1 SSME after detecting an over-temp condition. This caused an "abort-to-orbit" where the remaining two SSMEs were throttled up to 109% and the two OMS engines were fired to provide enough energy to reach orbit. While this procedure was underway, a second sensor indicated one of the remaining SSMEs was overheating and was about to be shut down by the GPCs. Had this engine also failed, the orbiter would have attempted a "trans-atlantic-abort" to an emergency landing in Rota, Spain. Fortunately, a flight controller in Houston suspected the sensor was malfunctioning and ordered the crew to disable the sensor prior to the computer's commanding an engine shutdown. *Challenger* entered an initial orbit of 124 by 165 miles, that later was corrected to 196 by 207 miles. This flight carried the third Spacelab mission, which covered a broad range of experiments in plasma physics, astrophysics, solar astronomy, and materials processing.

Flight 20 (STS-27, 51-I) — This launch was planned originally for August 24, but was delayed due to thunderstorms and lightning in the pad areas. It was delayed further until August 27 to replace a failed General Purpose Computer (#5), and to inspect the engine ducts. The launch on August 27 was trouble-free. *Discovery* deployed three communications satellites. The Leasat-3 satellite which failed to activate after deployment on Flight 16 was retrieved, repaired, and successfully redeployed.

Flight 21 — (STS-28, 51-J) — After an FRF during September, *Atlantis* was devoted to a classified DoD mission during its first flight. The Air Force again issued a statement calling the flight "...successful...".

Flight 22 — (STS-30, 61-A) — *Challenger* carried the fourth Spacelab mission devoted to materials processing experiments, and the largest flight crew to date—eight. For the first time a ground station other than Mission Control in Houston controlled portions of the flight. This was the German Space Operations Center located in Oberpfaffenhofen, near Munich, Germany.

Flight 23 — (STS-31, 61-B) — This was the second night launch of the program. *Atlantis'* flight was highlighted by astronaut assembly of structures in orbit, and attendant study of extravehicular dynamics and human factors. These activities were filmed by high resolution IMAX cameras mounted in the payload bay. The mission also deployed three communications satellites. All three were boosted successfully into geosynchronous orbit by their Payload Assist Modules.

Flight 24 — (STS-32, 61-C) — Fresh from a major refit in Palmdale, *Columbia's* launch would be delayed several times from a planned date of December 18. On that date it was delayed because of the excess time needed to close out the aft compartment. On December 19, the count was halted at T-14 seconds due to an out-of-tolerance reading in the right SRB hydraulic system. Another launch attempt on January 6 was halted at T-31 seconds due to a problem with the fill and drain valve in the LO_2 system; the window ended before the problem could be resolved. On January 7 the launch team tried again, but marginal weather for an emergency return to KSC, plus bad weather at the emergency landing sites at Dakar and Moron forced a postponement. The January 9 planned launch date was delayed an extra day to permit removal of an obstruction in the SSME #2 liquid oxygen prevalve. On January 10 another launch attempt was made, but eventually called off due to heavy rains in the pad area. The actual lift-off on January 12 then was achieved without major incident. This record five attempts to get *Columbia* off the pad caused serious work-load problems at the various NASA centers and contractor sites. *Columbia* deployed a communications satellite, conducted experiments in infrared imaging, and acquired photos and spectral images of Halley's Comet. Florida Congressman Bill Nelson was among the crew of seven. Mission controller's decided to shorten the planned flight by one day to provide more processing time on the ground for the next flight of *Columbia*, bringing it in at KSC on January 16. The landing attempt had to be waved off on that day due to unfavorable weather at KSC, and waved off the next day as well. The mission was extended for one more day for a third attempt at KSC, but when that was cancelled, *Columbia* landed at Edwards instead.

Flight 25 — (STS-33, 51-L) — January 28, 1986, the orbiter *Challenger* exploded 73 seconds after launch.

THE ACCIDENT

The launch date of Flight 25 (STS-33/51-L) was postponed three times and scrubbed once from the planned date of January 22, 1986. The first postponement was announced on December 23, 1985, adding an extra day (until January 23) to accommodate the final integrated simulation that had slipped one day due to the late launch of Flight 24. On January 22, 1986, the date was slipped to January 26, primarily because of Kennedy work requirements. The third postponement occurred on January 25 when forecasts indicated the weather would be unacceptable. The new launch date was set for January 27.

The launch attempt on January 27 began with fueling the external tank at 00:30 hours, Eastern Standard Time. The crew awakened at 05:07, and events proceeded normally with the crew strapped into the orbiter at 07:56. At 09:10, however, the countdown was halted when the ground crew reported a problem with the crew hatch exterior handle. By the time the hatch problem was resolved at 10:30, the Kennedy recovery site designated for a return-to-launch-site abort had exceeded the allowable velocity for crosswinds. The launch attempt was scrubbed at 12:35, and rescheduled for the following morning.

The weather on January 28 was forecast to be clear and cold, with temperatures dropping into the low twenties overnight. The management team directed engineers to assess the possible effects of temperatures on the launch. No critical issues were identified to management officials, and while evaluation continued, it was decided to proceed with the countdown and fueling of the ET.

A significant amount of ice had accumulated on launch pad 39B overnight, and it caused considerable concern for the launch team. In reaction, the ice inspection team was sent to the launch pad at 01:35, and returned to the launch control center at 03:00. After meeting to consider the team's report, the space shuttle program manager decided to continue the countdown, and scheduled another ice inspection for T-3 hours.

At 08:44 the ice team completed its second inspection. After hearing the team's report, the program manager decided to allow additional time for the ice to melt. He also decided to send the ice team to perform one final inspection during the scheduled hold at T-20 minutes. At this point,the launch had been delayed two hours past its original scheduled time of 09:38.

At 11:15 the final ice inspection was completed, and during the scheduled hold at T-9 minutes, the flight crew, and all members of the launch team, gave a "go" for launch. The final flight of *Challenger* began at 11:38:00.010, Eastern Standard Time, January 28, 1986.

From lift-off until telemetry from the shuttle was lost, no flight controller observed any indications of a problem. The shuttle's main engines throttled down to limit the maximum dynamic pressure, then throttled up to full thrust as expected. Voice communications were normal. The crew called to indicate the vehicle had begun its roll to head due east, and fifty-five seconds later, Mission Control informed the crew that the engines had throttled up successfully, and all other communications were satisfactory. Dick Scobee's acknowledgement of this call was the last voice communication from *Challenger*.

There were no alarms sounded in the cockpit, and the crew apparently had no indication of a problem before the rapid break-up of the vehicle. The first evidence of the accident came from live video coverage and when radar began tracking multiple targets. The flight dynamics officer in Houston confirmed to the flight director that "... RSO (Range Safety Officer) reports the vehicle has exploded...", and thirty seconds later added that the Air Force range safety officer had sent the destruct signal to the solid rocket boosters.

During the period of flight while the solid rocket boosters are thrusting, there are no survivable abort options; there was nothing that either the crew, or the flight controllers, could have done to avert the catastrophe.

A combined Coast Guard/NASA/Air Force/Navy search team spent the next three months searching the Atlantic for the remains of *Challenger* and her crew. Approximately 30% of the orbiter was recovered, including all three SSMEs, the forward fuselage with the crew module, the right inboard and outboard elevons, a large portion of the right wing, the lower portion of the vertical stabilizer, three rudder/speed brake panels, and portions of the mid-fuselage. The debris was evaluated by the NASA and NTSB in typical accident investigation fashion.

THE ANALYSIS

President Reagan, seeking to ensure a thorough and unbiased investigation of the *Challenger* accident, announced the formation of a Presidential Commission on February 3, 1986. The commission was chaired by William P. Rogers, a former Secretary of State under President Nixon, and other members including astronauts Neil A. Armstrong and Sally K. Ride; Robert B. Hotz, former editor-in-chief of *Aviation Week & Space Technology*; Brigadier General Charles Yeager, USAF (Retired); and several distinguished scientists and engineers including a Nobel laureate.

The report of the Commission concluded:

> "The consensus of the Commission and participating investigative agencies is that the loss of the Space Shuttle *Challenger* was caused by a failure in the joint between the two lower segments of the right Solid Rocket Motor. The specific failure was the destruction of the seals that are intended to prevent hot gases from leaking through the joint during the propellant burn of the rocket motor. The evidence assembled by the Commission indicates that no other element of the Space Shuttle system contributed to this failure."

More than 160 individuals were interviewed and more than 35 formal panel investigative sessions were held, generating almost 12,000 pages of transcript. Nearly 6,300 documents, totalling more than 122,000 pages, and hundreds of photographs were examined and made a part of the Commission's permanent record.

The series of events leading up to the destruction of *Challenger* was determined to have started with a leak in an O-ring seal in the right SRB. This leak produced severe heating of the liquid hydrogen tank and lower ET/SRB sway strut, causing the latter to fail. Upon failure of the sway strut, the SRB pivoted on the forward thrust strut, causing damage to the orbiter's right wing and to the liquid oxygen tank on the top of the ET. The asymmetrical thrust caused by this swaying of the right SRB was more than the thrust vector control system was designed to counter. Damage to the LO_2 tank, and the extreme heating of the LH_2 tank by the leak in the right SRB, eventually caused the ET to explode. All fractures and material failures examined on the orbiter, with the exception of the main engines, were the result of overload forces, and they exhibited no evidence of internal burn damage or exposure to explosive forces. This indicated the destruction of the orbiter occurred predominantly from aerodynamic and inertial forces that exceeded design limits. Additionally, chemical analysis indicated that the right side of the orbiter was sprayed by hot propellant gases exhausting from the hole in the inboard circumference of the right SRB. Evaluation of the SSMEs showed extensive internal thermal damage to the engines

Rockwell International via Erik Simonsen

"Discovery" roll-out on October 16, 1983 at Rockwell's Palmdale facility. For the first time, insulation "blankets" were used instead of white tiles on the forward fuselage.

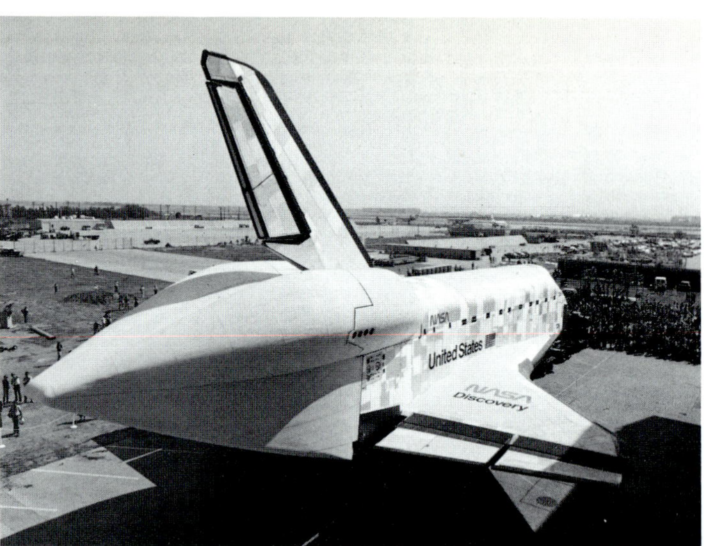

Rockwell International via Erik Simonsen

"Discovery", during October 16, 1983 roll-out. Panel just below OMS pod is where the tail service masts connect to provide power, LO_2, LH_2, and purging gases.

as a consequence of an oxygen-rich shutdown. The supply of hydrogen fuel to the main engines would have been abruptly terminated when the liquid hydrogen tank in the ET disintegrated after being super-heated by the exhaust gases escaping from the leak in the right SRB. The crew module wreckage was found submerged in about 90 ft. of ocean water concentrated in an area of about 20 ft. by 80 ft. Portions of the forward fuselage outer shell structure were found among the pieces of the crew module. There was no evidence of internal explosion, heat or fire damage on the forward fuselage/crew module. The crew module was disintegrated, with the heaviest fragmentation and crash damage on the left side. The fractures were typical of overload forces resulting from the impact with the ocean. The consistency of damage to the left side of the outer fuselage and the crew module indicated these structures remained attached to each other and intact until impact with the water. The entire ET range safety system was recovered intact, and the SRBs were observed to have destructed upon command, thus ruling out any possibility that the RSS malfunctioned and caused the accident.

Following the completion of the accident investigation, the remains of *Challenger* were sealed permanently in two surplus *Minuteman* ICBM test silos on the Cape Canaveral Air Force Station.

O-ring anomalies had been detected to varying degrees on 12 previous flights. Erosion of either the primary or secondary O-rings had been seen on Flights 2, 10, 11, 12, 15, 16, 17, 18, 20, 22, 23 and 24. A more serious problem, the actual blow-by of exhaust gases past an O-ring had occurred on Flights 11, 12, 15, 16, 17, 18, 22, 23 and 24. All of these anomalies were recorded upon occurrence, and this data was known at the Flight Readiness Review for Flight 25. Failure analysis conducted as early as 1979 on the shuttle system had concluded that one in fifty flights would encounter a catastrophic accident during ascent, and one in 100 would fail to land successfully. The failure analysis have been updated repeatedly, and always reaches much the same conclusions.

NASA

"Discovery", minus OMS pods, SSMEs, and numerous insulation tiles, in the VAB high bay 2 area.

RETHINKING ABORTS AND ESCAPES

Between 1973 and 1983, first stage (while the SRBs are thrusting) abort provisions were assessed many times by all levels of NASA and contractor management. Many methods of saving the orbiter and/or crew have been proposed, and rejected. Any method of ejection, or separable crew modules were continually rejected due to their limited utility and significant cost and weight impacts. Because of these factors, NASA adopted the philosophy that the reliability of the first stage (SRBs) must be assured, and that design and testing must preclude time critical failures that would require emergency action before the normal solid rocket booster burnout.

If a problem arose that required the orbiter to get away from the solid rocket boosters, the separation would have to be performed extremely quickly. Time is of the essence for two reasons: first, as 51-L demonstrated, if a problem develops in the SRBs, it can escalate very quickly; second, the ascent trajectory is carefully designed to control aerodynamic loads on the vehicle; a very small deviation from the programmed path will produce excessive loads, so if the vehicle begins to diverge from its path there is very little time (seconds) before structural breakup will occur.

The normal separation sequence to free the orbiter from the rest of the system takes 18 seconds, far too long to be of use during a first-stage emergency. Therefore a capability called "fast-separation" was built into the flight software for use at any time. Fast-separation bypasses or reduces the normal built-in delays in order to achieve separation in approximately three seconds. Some risk was accepted to obtain this contingency capability. Unfortunately, subsequent analysis has shown that if it is attempted while the SRBs are thrusting, the orbiter will "hang-up" on its aft attach points and pitch violently, with the probable destruction of the vehicle. Therefore fast separation does not provide a meaningful way to escape unless some form of SRB thrust-termination system is implemented, and this was rejected for sound technical reasons during 1973.

The current concept of fast-separation does have some uses. Contingency aborts resulting from the loss of two or three main engines soon after SRB separation will require quick separation from the ET so the orbiter can attain an entry attitude quickly. Unfortunately, all contingency aborts of this type culminate in water impact, and it is extremely unlikely that the orbiter, or the crew, could survive a ditching. It is desirable therefore to have available to the crew a reasonable "bail-out" option during controlled, gliding flight.

THE RECOMMENDATIONS

Although the Commission found the 51-L accident to be a direct result of the SRB seal failure, numerous other areas of concern were uncovered during the investigation. These areas included everything from the orbiter braking system, to a decision making process described as "flawed". It was found that in its efforts to produce an "operational" system, NASA had abandoned many of the procedures that had made it successful during its first 25 years.

The Commission provided NASA with nine major recommendations to help ensure a safe return to flight:

1. Redesign the SRB joint and seal. The recertification process should more closely resemble flight conditions and loads. Alternate SRB designs should be evaluated. Procedures for assembly and inspection of SRBs prior to launch operations should be evaluated and refined.
2. Review the shuttle management structure. An effort should be taken to consolidate decision making and review processes into a shuttle program organization instead of as an individual center effort. The earlier NASA practice of utilizing astronauts in management positions should be reinstituted. An independent flight safety panel should be established.
3. A complete review of all safety critical items should be undertaken and the procedures for granting waivers should be tightened.
4. NASA should establish an Office of Safety, Reliability and Quality Assurance reporting directly to the NASA Administrator.
5. NASA should improve communications between centers, and the designated flight commander (or his representative) should participate in the Flight Readiness Review process.
6. Actions should be taken to improve landing safety, especially the orbiter brakes and nosewheel steering. A second Shuttle Carrier Aircraft should be purchased.
7. Efforts should be made to provide a crew escape system for use during controlled, gliding flight.
8. NASA must establish a flight rate that is consistent with its resources. The nation's reliance on a single launch system should be avoided in the future.
9. Installation, test and maintenance procedures must be especially rigorous for critical items. An improved system for tracking maintenance intensive items should be implemented.

During the thirty-two month stand-down after the *Challenger* accident, NASA has made great strides towards meeting most of these recommendations. By September 1, 1988, 76 mandatory orbiter modifications had been completed, including the crew escape system, increasing the thermal protection on the chin panel, new brakes, and recertifying the 17-inch propellant disconnects between the orbiter and the ET. One-hundred and eighty-five ground system modifications also have been accomplished. But by far the most extensive work has been to the Solid Rocket Boosters.

SRB RECERTIFICATION

A comprehensive test program was developed by Mortion Thiokol to evaluate and verify all changes incorporated into the redesigned solid rocket boosters. The tests were designed to thoroughly evaluate design changes under all relevant environments and loading conditions. Each static test was full duration (122 seconds) with the booster in a horizontal position. Each booster was fitted with 400 to 600 channels of instrumentation to measure pressure, deflection, thrust, strain, temperature and other conditions.

A booster (Engineering Test Motor [ETM-1A]) with 51-L configuration hardware was fired on May 27, 1987. The two forward field joints incorporated the original Viton O-rings and external graphite reinforcing bands. The aft field joint had one Viton and one silicone O-ring. All three joints were equipped with wrap-around electrical joint heaters. In addition to evaluating new O-ring seal material in unmodified hardware, ETM-1A was used to test the effectiveness of the external graphite composite stiffener rings in reducing joint rotation. It also was used to evalute the performance of field joint heaters, measure baseline hardware joint deflections in both the field and case-to-nozzle joints, and define the effectiveness of new nozzle nose-inlet rings.

The first test of the redesigned configuration was Demonstration Motor Eight (DM-8). The DM-8 static test on August 30, 1987 was used to evalute the performance of major design features of the redesigned motor, including the prototype bonded J-seal insulation design. It also was used to evaluate the redesigned structural backup nozzle outer boot ring and to evaluate the effects of the External Tank attachment ring on the redesigned hardware. The capability of the joint heaters to maintain seal temperatures of 75° F also was evaluated. DM-9, the second full-scale test of the redesigned motor was fired on December 23, 1987 under ambient temperatures of 20° F. The test was designed to further study the performance of the major redesign features, and further evaluate the joint heaters.

On April 20, 1988, Qualification Motor Six (QM-6) started the process of recertifying the SRBs for manned flight. The center field joint was assembled with an intentional manufacturing-type flaw in the surface of the bonded insulation J-seal. The defect was designed to allow hot gasses to penetrate the bonded insulation as far as the capture O-ring. Also incorporated into QM-6 was an intentional flaw in the case-to-nozzle joint which allowed gasses to penetrate the bonded insulation to the new wiper O-ring. QM-6 and all subsequent boosters included the structural backup configuration outer boot ring. The QM-7 static test on June 14, 1988 was used to evalute the redesign without intentional manufacturing or assembly flaws in the joints. Three hydraulically actuated struts, programmed to simulate ignition, liftoff, and flight loads, were attached externally to the QM-7 test article.

Production Verification Motor One (PV-1) was fired on August 18, 1988 and was the final qualification test prior to the launch of STS-26R. PV-1 was the most seriously flawed motor ever tested, containing 14 distinct flaws in all but one of the major joints. Post-test disassembly and inspection demonstrated that all joints performed as predicted. One final cold weather test was performed during December 1988, after the STS-26R and STS-27R missions.

TECHNICAL DESCRIPTION:

The National Space Transportation System consists of three components: the orbiter, the External Tank (ET) and the Solid Rocket Boosters (SRB). Of these, only the External Tank is not recovered and reused (proposals are being considered calling for these components to be carried into orbit and used as part of an orbiting laboratory).

ORBITER DESCRIPTION

The orbiter is a double-delta winged reentry vehicle capable of carrying both passengers and cargo to low-earth orbit and back to a controlled gliding landing. The spacecraft consists of a crew module pressure vessel contained in the forward fuselage, wings, payload bay doors, mid-fuselage, aft-fuselage and a vertical stabilizer. The orbiter is constructed of aluminum alloys covered by reusable surface insulation in tile and blanket form. Some composites are used for high-temperature areas on the wing and vertical tail leading edges and nose.

FORWARD FUSELAGE

The forward fuselage is made using conventional aircraft manufacturing techniques and contains the pressurized crew module. The fuselage structure is built of type 2024 aluminum alloy stringer panels, frames and bulkheads. The panels are made of stretched-formed skins with riveted stringers three to five inches apart. Frames are riveted to the stringer panels, with 30 to 36 in. spacing between each. A formed upper fuselage section and a machined lower fuselage section make up the forward bulkhead. The bulkhead provides structure for the nose section, which contains large machined beams and struts. The nose landing gear well structure consists of two support beams, two upper closeout webs, drag-link support struts, hydraulic cylinder attachment points, and the landing gear door fittings. The nose landing gear doors are constructed of aluminum alloy honeycomb and the left-hand door is wider than the right-hand door. The forward Reaction Control System module is built of aluminum and attaches to the forward fuselage at 16 fittings. The nosecap is made of reinforced carbon-carbon for protection from peak reentry heating.

CREW MODULE

The crew module is a pressurized compartment housing flight control stations, living quarters and equipment bays. It has a volume of 2,525 cubic ft. and contains three levels. The upper level is the flight deck and provides seating for up to four crewmembers. The mid-deck contains passenger seating, a living area, and an airlock into the payload bay, along with avionics equipment compartments. The lower deck contains the environmental control equipment and is accessible through removable floor panels on the mid-deck. The crew module basically is conical, and is composed of 2219 aluminum alloy machined to provide stringers, frame lands and local reinforcement. It is attached to the lower and upper forward fuselage by four vibration and shock absorbing mounts.

FLIGHT DECK

The flight deck is designed with the standard pilot and co-pilot seating arrangement, allowing the vehicle to be flown from either station, or by one crewmember. Seating is provided for four crewmembers on the flight deck. During the Approach and Landing Test (ALT) flights *Enterprise* was equipped with Lockheed (modified A-12/SR-71 units) zero-zero ejection seats for the two pilot positions, and no other seating was provided. During the orbital flight test phase (STS-1 through STS-4), *Columbia* also was equipped with Lockheed zero-zero ejection seats for use on the launch pad, and during landing operations. These seats were disabled after STS-4, and removed during a refit at Palmdale after STS-5. No crew escape system was provided during STS-5 through STS-33.

After the loss of *Challenger*, the issue of crew escape was reexamined, and after again rejecting various proposals for ejection seats, escape pods, etc., two different options were proposed: a telescopic slide-pole, and a tractor rocket system. The tractor rocket system provided small rockets that could be attached to the individual parachutes and fired from the crew access hatch to carry the crew member clear of the orbiter.

The chosen system is a 9.8 ft. long telescopic pole that is spring loaded to extend from the crew access hatch. During evacuation, crew members slide down the pole with special attachments on their parachute harnesses. The pole curves sharply downward and slightly aft, directing the crew member under the orbiter's left wing. Both methods were considered useful only below 200 knots and 20,000 ft. during controlled gliding flight. This escape system was tested on a Lockheed C-141B and a Convair C-131A aircraft during the spring of 1988, and was found to provide a crew of seven with evacuation times of about 100 seconds. William A. Chandler, crew system escape manager at the Johnson Space Center, noted that although both the slide-pole and tractor rocket systems provide adequate escape clearance for the crew, the slide-pole system is considered safer and will prove more cost effective over the life of the orbiters. Tractor rockets would have necessitated carrying pyrotechnics in the orbiter cabin, and would have required additional support during the turnaround periods between flights. Modifications required to install the new system in *Discovery* were completed on April 15, 1988, and the system was operational in time for the launch of STS-26R.

Over 2,000 displays and controls are located on the flight deck instrument panels. Additional functions are controlled through keyboards interfacing directly with the onboard IBM General Purpose Computers (GPC). Conventional stick controllers are used for flight controls, but there are no throttles in the conventional sense, this function being performed by the GPCs. A stick-type controller also is provided at the aft crew station for control of the Remote Manipulator System (RMS).

Aft of the pilot's seat is the mission station. Located on the aft bulkhead wall are displays and controls for payloads contained in the payload bay. A dedicated caution-and-warning display is provided for payloads and orbiter systems that do not require interaction with the flight crew.

The six windshields on the flight deck provide pilot visibility, and are the largest pieces of optical quality glass ever produced for "see-through" viewing. Each of the outer windshields on the forward fuselage is 42 in. diagonal, and the two inner panels are 35 in. diagonal. The innermost pane in the crew module is constructed of tempered aluminosilicate glass to withstand the cabin pressure with a high degree of reliability, and is 0.625 in. thick. The outer surface of this pane is coated with an infrared-reflector that transmits the visable spectrum. A beryllium and aluminum retainer is used to attach the inner pane to the window frame.

The center transparent pane in the crew module is constructed of low expansion fused silica glass. This material is used because of its high optical quality and excellent thermal shock resistance. It is 1.3 in. thick. The surfaces of this pane are coated with a high efficiency anti-reflection coating to improve visible light transmission. These windows must successfully withstand a stress pressure of 8,600 lbs. per square inch in a flexure test performed at 240°. The windshields are held in the crew module by glass retainers. In the innermost panel (pressure panel), grooves are provided in the glass retainer for redundant seals which utilize a berrylium and aluminum flat glass retainer to attach the inner pane to the window frame. No sealing or bonding compounds are used.

The outer panes are installed in the forward fuselage, and are 0.625 in. thick. The exterior of these panes is uncoated, but the interior uses the same anti-reflective coating as the center pane. The outer surface of the window is designed for a reentry temperature of 900°.

MID-DECK

The mid-deck contains crew quarters and equipment storage. Normal crew size is seven, although the sleeping and some storage areas can be replaced by additional seating if required, providing a maximum crew size of ten for short duration missions. Access to the flight deck is provided by a 26 x 28 in. hatch. A ladder attached to the portside interdeck opening allows access to the flight deck while on the ground. The outside crew hatch is located on the port side of the mid-deck, and the airlock to the payload bay is located in the aft bulkhead. Also located on the portside wall of the mid-deck are lockers for stowage, a waste management facility (toilet) and personal hygiene station, privacy screen, and galley. Located on the forward wall are additional stowage space, a work or dining table and lockers for experiments. The starboard wall contains three drawer-like sleep stations, a vertical sleep restraint station and modular stowage space.

The side hatch in the mid-deck is utilized for normal ingress/egress of the crew, and may be operated from within the crew module, or externally. The side hatch is 40 in. in diameter, contains a 10 in. diameter window in the center and opens outward. The window is constructed of three panes of glass of the same type as the orbiter windscreens.

EQUIPMENT BAY

Located below the mid-deck is an equipment bay for storage of expended lithium hydroxide canisters (used in purging contaminants from the breathing air) and trash. Access to the storage area is provided by a hatch in the mid-deck floor. Environmental Control System equipment may be accessed through removable floor panels. Forward of the mid-deck is an unpressurized area containing two avionics bays. These bays contain racks for electrical and navigation equipment and can not be accessed by the crew.

MID-FUSELAGE

The mid-fuselage, built by General Dynamics, contains the payload bay and connects with the forward fuselage, aft fuselage, wings and payload bay doors. It is 60 ft. long, 17 ft. wide, 13 ft. high and weighs approximately 13,500 lbs. It contains 12 main frame assemblies that provide stabilization of the structure. Each frame consists of machined vertical sides and horizontal braces constructed of boron/aluminum tube trusses. The upper portion consists of door and sill longerons that absorb the bending loads of the vehicle and support paylods contained in the payload bay. The mid-fuselage contains thirteen Inconel-718 hinges for the payload bay doors that provide an opening for payload installation and deployment. The doors are hinged at the sides and latched forward and aft at the vehicle centerline.

Because of additional analysis of actual flight data concerning descent-stress thermal gradient loads, torsional straps were added to tie all the lower mid-fuselage stringers in bays 1 through 11 together in a manner similar to a box section. This eliminates rotational (torsional) tendencies, and provides additional positive margins of safety. Also, based on findings from this analysis, room-temperature vulcanizing silicone rubber material was bonded to the lower mid-fuselage from bays 4 through 11 to act as a heat sink, distributing temperatures evenly across the mid-fuselage bottom.

The payload bay doors also serve as a strongback for the spacecraft radiator panels. The two forward radiator panels can be unlatched and tilted to allow heat to be radiated from both the front and back sides of the panels; the two fixed aft panels dissipate heat only from the outer side. Each payload bay door is 60 ft. long and has a mean chord of approximately 10 ft. The payload bay doors are constructed of a graphite-epoxy/Nomex composite, weigh 1,632 lbs. each, and are 23% lighter than an equivalent aluminum honeycomb sandwich. Payloads and payload carriers are attached to structural supports along the orbiter's mid-fuselage longerons.

Spar Aerospace, Limited, of Toronto, Canada, is the prime contractor for the Remote Manipulator Subsystem (RMS). The basic RMS configuration consists of an manipulator arm, an RMS display and control panel (furnished by CAE Electronics, Ltd.) and an interface to the orbiter computers. Most missions carry one arm mounted ont he left main longeron of the payload bay upper wall. Provisions exist to mount a similar device along the right side, and both may be carried on a single mission if requirements dictate, although due to software limitations, only one may be used at a time. Each manipulator arm is capable of handling payloads up to 60 ft. long and weighing 60,000 lbs.

The manipulator arm is 50 ft., 3 in., in length, 15 in. in diameter and has six degrees of freedom. The upper and lower arm booms are constructed of graphite-epoxy composite and weigh 93 lbs. each. The joints and electronic housings are constructed of aluminum alloy. Six flood lights, three mounted on each side of the payload bay, provide lighting during RMS operations.

A complete RCA Astro-Electronics closed circuit TV system includes a color camera in the crew compartment and several black-and-white or color cameras in the payload bay and on the manipulator arms. These facilitate payload handling and provide TV coverage for engineers and the general public on earth.

WINGS

The blended delta wings of the orbiter provide aerodynamic lift and control of the vehicle during atmospheric flight. The main wing assemblies are constructed by Grumman Aerospace. They consist of a wing glove, an intermediate section that contains wells for the landing gear, torque box, forward spar for mounting of the leading edge thermal protection, and elevons. The wings have a modified NACA-0010 section, with an 81° sweep on the inner leading edge, and 45° on the outer. The trailing edge has a 3°31' dihedral. Each of the main wing assemblies are 60 ft. long and have a maximum thickness of five feet at the fuselage intersection.

The forward wing box is of conventional design with a corrugated spar web and aluminum truss ribs with sheet metal caps. The upper and lower wing skin panels are stiffened aluminum.

The intermediate wing section consists of the same conventional design of aluminum multi-ribs and tube truss. The upper and lower skins are aluminum honeycomb construction. A portion of the lower wing skin panel is made up of the main landing gear door. The intermediate section houses the main landing gear compartment, and absorbs a portion of the main landing gear loads. A structural rib provides support for the outboard main landing gear door hinges, and the main gear trunnion and drag link. The support for the inboard main gear trunnion and drag link attachment is provided by the mid-fuselage.

The main landing gear doors are constructed of conventional aluminum with machined aluminum hinge beams and hinges. The remainder of the door is conventional aluminum beams and multi-stringers with stiffened aluminum skin. A recessed area in the door is provided for tire clearance.

The wing torque box area incorporates a conventional multi-rib (11) truss arrangement with four aluminum spars. The spars are of a corrugated aluminum configuration to minimize heat loads. The forward closeout beam is of aluminum honeycomb and provides the attachment of the leading edge thermal protection system. The rear spar provides the attachment interfaces for the elevons, hinged upper seal panels and associated hydraulic and electrical systems. The upper and lower wing skin panels are of aluminum stiffened skin covered with thermal protection tiles and blankets.

The elevons provide flight control during atmospheric flight. The elevons are of conventional aluminum multi-rib and beam construction with aluminum honeycomb skins. Each elevon is divided into two segments, and each segment is supported by three hinges. Attachments for the flight control systems are provided along the forward extremity of the elevons, and all hinge moments are at these points. The upper leading edge of each elevon panel incorporates rub strips of titanium construction. Hinged panels on the upper wing surface, of titanium and Inconel sandwich, are used to seal the wing/elevon gap, and are not covered by the thermal protection system.

Attachment of the wing to the fuselage is accomplished with a tension bolt splice along the upper surface. A shear bolt splice along the lower surface in the area of the fuselage carry-through completes the fuselage attachment interface.

Before the wings for *Discovery* and *Atlantis* were manufactured, a weight reduction program was instituted that resulted in a redesign of certain areas of the wing structure. An assessment of wing air loads from actual flight data indicated greater loads on the wing structure than predicted. To maintain positive margins of safety during ascent, structural modifications were incorporated during the 51-L stand-down into OV-103 and OV-104. Since *Columbia* used a wing of older, more conservative design, modifications were not deemed necessary. The wing for OV-105 will be modified during assembly.

AFT-FUSELAGE

The aft-fuselage consists of an outer shell, thrust structure, and internal secondary structure. It provides the primary support for the spacecraft's Main Propulsion System, Orbital Maneuvering System pods and vertical stabilizer. The structure is 18 ft. long, 22 ft. wide and 20 ft. high.

The aft fuselage provides the load path to the mid-fuselage main longerons, main wing spar continuity across the forward bulkhead of the aft fuselage, structural support for the body flap, structural housing around all internal systems for protection from the operational environment (pressure, thermal, and acoustic), and for controlled internal pressure venting during flight.

The forward bulkhead closes off the aft fuselage from the mid-fuselage and is composed of machined and beaded aluminum sheet segments. The upper portion of the bulkhead attaches the front spar of the vertical stabilizer.

The internal thrust structure supports the Space Shuttle Main Engines (SSME). This structure includes the SSME load reaction truss structure, engine interface fittings and the SSME gimbal actuator support structure. The upper thrust structure of the aft-fuselage is of integral machined aluminum construction with aluminum frames, with the exception of the fin support frame, which is titanium. The skin panels are integrally machined aluminum and attach to each side of the vertical stabilizer to react to drag and torsion loading.

The outer shell of the aft-fuselage is constructed of integral machined aluminum. Various penetrations are provided in the outer shell for access to installed systems. The secondary structure is of conventional aluminum construction, except for the utilization of some titanium and fiberglass for thermal isloation of selected equipment. The structure consists of brackets, built-up webs, truss members and machined fittings as required by subsystem loading and support constraints. Certain system components, such as the avionics shelves, are shock-mounted to the secondary structure. Support provisions also are included for the auxiliary power units, hydraulics, environmental control/life support systems, and electrical wire runs. The heat shield provides the close-out of the aft base area of the aft-fuselage and consists of machined aluminum flat panels and fiberglass dome panels.

BODY FLAP

Pitch control during atmospheric flight is provided by the body flap. The flap also shields the Main Engines from reentry thermal conditions during descent. It is of conventional aluminum multi-rib construction with upper and lower skins of aluminum honeycomb. The body flap is attached to the aft fuselage by four rotary actuators.

VERTICAL STABILIZER

The vertical stabilizer consists of a structural fin surface, the rudder/speedbrake (the rudder splits into two halves for speedbrake control), and systems required to position the rudder/speedbrake control surfaces. It has a leading edge sweep of 45°.

The vertical stabilizer structure is built by Fairchild Republic and consists of aluminum integral ribs, webs, stringers and machined aluminum spars to make a torque box for primary load carrying. The fin lower trailing edge area that houses the rudder/speedbrake power drive unit has aluminum honeycomb skins. Attachment of the stabilizer is accomplished by two tension tie bolts at the root of the front spar of the tail to the forward bulkhead of the aft fuselage and by eight shear bolts at the root of the stabilizer rear spar to the upper structure portion of the aft-fuselage.

The rudder/speedbrake control surface consists of conventional aluminum ribs and spars with aluminum honeycomb skin panels for the primary load carrying members, and is attached through rotating hinge parts to the vertical tail fin. An Inconel honeycomb aerodynamic seal is employed between the rudder/speedbrake and the fin, and is not covered by the thermal protection system.

Mission requirements call for a locked rudder/speedbrake during ascent, orbit and high-speed reentry. Speed brake control is provided from Mach 10 to Mach 5. Below Mach 5, speed brake and rudder controls are provided as required.

MAIN ENGINES

The three Space Shuttle Main Engines (SSME) are built by the Rocketdyne Division of Rockwell International, and are located in the aft section of the aft-fuselage. The SSMEs provide thrust in conjunction with the two Solid Rocket Boosters during the ascent phase of shuttle operations.

Liquid hydrogen (LH_2) and liquid oxygen (LO_2) are used as fuel and oxidizer, respectively, and are supplied from the external tank. The LH_2 and LO_2 tanks are pressurized with gaseous helium while on the launch pad, which forces the propellants from the ET through 17 in. feedlines in the aft underside of the orbiter. Inside the orbiter, these 17 in. feedlines branch into separate 12 in. feedlines to each SSME. After engine start, the SSMEs provide the pressurization for the propellant tanks by extracting the respective vaporized propellant from the engines. The propellants are further pressurized at each engine interface by high-speed, high-capacity turbopumps.

Rockwell International via Erik Simonsen

"Discovery" during roll-out on October 16, 1983 at Palmdale. The small notch just forward of and below the cockpit window is one of the star-trackers used for stellar navigation (a second is located just above it).

Each engine combines the merits of high-chamber pressure operation, a high-performance, bell-shaped nozzle, and a regeneratively cooled thrust chamber for maximum performance and long life. In regenerative cooling, fuel is moved through passages in the thrust chamber walls, then returned to the injector to be burned with the balance of the propellant. This allows higher combustion temperatures than would be possible without cooling the combustion chamber walls.

Each SSME consumes 1,035 lbs. of propellant per second (889 lbs. of LO_2, 146 lbs. of LH_2) and burns at 3,000 psia. Each engine develops 375,000 lbs. of thrust at sea level, and 470,000 lbs. in a vacuum. A specific impulse of 455 seconds is achieved. The mixture ratio of liquid oxygen to liquid hydrogen is 6 to 1. Each engine currently is throttleable from 50% to 109% of rated thrust, although the upper limit will be increased during 1989 to 120%. Each SSME is designed for 7.5 hours of operation, or roughly 55 flights at 8 minutes per flight, with a major overhaul accomplished every 9 flights.

Each SSME has an engine control computer that monitors parameters such as pressure and temperature, and automatically adjusts engine operation for the required thrust and mixture control. The system maintains a record of engine operating history for maintenance purposes, and relays real-time data to ground based controllers through a 60k bit/second telemetry link. The engines are gimbaled by hydraulic actuators for thrust vector control during ascent. Maximum gimballing capability is plus/minus 10.5° in pitch, and plus/minus 8.5° in yaw.

REACTION CONTROL SYSTEM

The Reaction Control System (RCS) has 38 bipropellant primary thrusters and six vernier thrusters to provide attitude control and three-axis translation during the orbit insertion, on-orbit, and reentry phases of flight. The RCS is used as the primary flight control when the vehicle is above 70,000 ft. The RCS thrusters are manufactured by the Marquardt Division of CCI Corporation. The primary thrusters are of type R-40A, and the vernier are R-1E-3 models.

The RCS consists of three propulsion modules, one in the forward fuselage, and one in each of the Orbital Maneuvering System (OMS) pods. All modules are used during external tank separation, orbital insertion and orbital maneuvers. Only the aft RCS modules are used for reentry attitude control.

Each of these RCS modules consists of two helium tanks that pressurize an oxidizer and fuel tank. The forward module contains 14 primary and 2 vernier thrusters. Each of the aft RCS modules have 12 primary and 2 vernier thrusters. The RCS propellants are Monomethyl Hydrazine (MMH) as the fuel and Nitrogen Tetroxide (N_2O_4) as the oxidizer. The design mixture ratio of 1.6 to 1 (oxidizer weight to fuel weight) permits the use of identical tanks for both fuel and oxidizer. The capacity of each tank is 1,488 lbs. of N_2O_4 or 930 lbs. of MMH. Since these substances are hypergolic in nature, no ignition system is required. The primary thrusters produce 870 lbs. of thrust in a vacuum, and a specific impulse of 289 seconds. Each of these thrusters is designed to be reusable for 100 missions, and is capable of sustaining 50,000 starts and 20,000 seconds of cumulative firing. The vernier thrusters produce 25 lbs. of thrust and a specific impulse of 228 seconds. Each of these is also designed for 100 missions, capable of being started 500,000 times for a total of 125,000 seconds of cumulative firing.

After the 51-L accident, the 64 valves operated by electric motors in the OMS and RCS were modified to incorporate a "sniff" line for each valve to permit monitoring for propellants in the electrical portion of the valves during ground operations. This new line reduces the probability of floating particles in the electrical microswitch portion of each valve, which could effect the operation of position indicators used for display and telemetry.

ORBITAL MANEUVERING SYSTEM

The Orbital Maneuvering Subsystem (OMS) provides the thrust to perform orbital insertion, orbit circularization, orbit transfer, redezvous and deorbit. The system consists of two pods, one on each side of the upper aft fuselage flanking the vertical stabilizer of the orbiter, that also contain the aft reaction control system. The pods and propellant tanks are fabricated by McDonnell Douglas Astronautics, and the OMS engines are manufactured by AeroJet Liquid Rocket Company. Graphite epoxy is used for external skins and for most bulkheads, ribs and stringers.

NASA via Erik Simonsen

"Discovery" moments before touching down on runway 17 at Edwards AFB following STS-14 on September 5, 1984. This was "Discovery's" first flight and three satellites were deployed. The payload of 47,500 lbs. set a shuttle record.

Each pod contains a high-pressure helium storage bottle, four tank pressure regulators and controls, a fuel tank, an oxidizer tank, and a pressure fed regeneratively cooled rocket engine. A maximum of 4,505 lbs. of monomethyl hydrazine (MMH) fuel, and 7,433 lbs. of nitrogen tetroxide (N_2O_4) oxidizer can be carried in each pod. The fuel and oxidizer used are the same as in the RCS.

Each engine produces 26,700 lbs. of thrust in a vacuum and specific impulse of 313 seconds at a chamber pressure of 125 psia. The fuel to oxidizer ratio is 1.65 to 1. Each engine is designed to be reusable for 100 missions, and capable of sustaining 1,000 starts and 15 hours of cumulative firing.

The integral OMS tankage is sized to provide propellant capacity for a change in velocity of 1,000 ft. per second at maximum gross weight. A portion of this velocity change is used during ascent. Depending on the payload in the orbiter, up to three OMS propellant kits may be carried in the payload bay, each kit providing an additional 500 ft. per second velocity capability.

The OMS engines are gimbaled by electro-mechanical actuators during firing for thrust vector control. The gimbal actuation system provides multi-axis gimballing of plus/minus 8 degrees.

POWER SYSTEMS

Electrical power is generated by three fuel cells that use cryogenically stored hydrogen and oxygen reactants. Each fuel cell is connected to one of three independent electrical buses. During peak and average power loads, all three fuel cells and buses are used; during minimum power loads, only two fuel cells are used, but they are interconnected to all three buses. The third fuel cell is put into a standby mode, and can be brought back online instantly. Excess heat from the fuel cells is transferred to a freon cooling loop through heat exchangers. The combined fuel cell system provides 14kW continuous/24kW peak at 28VDC. Each fuel cell has a life expectancy of 5,000 hours.

An extensive fuel cell redesign effort was undertaken during the stand-down caused by the 51-L accident. During this redesign numerous components of the fuel cells were improved in an effort to increase the reliability and maintainability of the system. Among the changes: End-cell heaters on each fuel cell powerplant were deleted because of potential electrical failures and replaced with freon coolant loop passages; the hydrogen pump and water separator in each fuel cell were improved to minimize excessive hydrogen gas entrained in the powerplant product water; several new sensors were added to provide more visibility into possible hydrogen pump overload and thermal conditions; and the product water from all three fuel cells now flows to a single water relief control panel to simplify ECLSS operations.

The Auxiliary Power Unit (APU) generates mechanical shaft power to drive a pump which produces pressure for the orbiter's hydraulic system. There are three separate Sunstrand APUs, and three 3,000 psi hydraulic pumps, one for each of the triple redundant hydraulic systems. The APU systems are located in the aft fuselage and consist of a fuel tank, a fuel feed system, an APU and controller, an exhaust duct and a lube oil cooling system. The APU fuel tanks are mounted to the sides of the aft fuselage and contain 100 gals. of hydrazine (N_2H_4) fuel. The fuel is fed to the APU under pressure, and provides for 81 minutes of operation. Each APU is rated at 135 hp at normal speed, sea level, and weighs just under 46 lbs. including its controller.

The APUs that have been used to date have a limited life, each unit being refurbished after 25 hrs. of operation because of cracks in the turbine housing. Improved APUs are scheduled for delivery during late 1988. A new turbine housing increases the housing life to 75 hrs. of operations (roughly 50 missions). The new APUs also incorporate a new gas generator valve module that deletes the requirement for a water spray system that previously was used for each APU upon shutdown. The deletion of the water spray system results in a weight savings of 150 lbs. per orbiter. Upon delivery of the new design APUs, the older, life-limited APUs will be refurbished to the new configuration.

ENVIRONMENTAL CONTROL

The Environmental Control and Life Support system is made up of three subsystems: the atmosphere revitalization subsystem which controls the atmospheric environment for occupants and the thermal environment for electronics; the food, water and waste subsystem to provide hygiene and other life support functions; and the active thermal control subsystem which maintains components within specified temperature limits and rejects excess heat via the payload bay door radiator panels.

THERMAL PROTECTION SYSTEM

The Thermal Protection System (TPS) protects the basically aluminum orbiter during the atmospheric portion of flight. The peak heating rates and longest exposure occur on high cross-range maneuvers during reentry when equilibrium surface temperatures range from 3,000° at stagnation points on the nose and leading edges of the wing and tail, down to 600° on leeward surfaces. The TPS is composed of two types of reusable surface insulation (RSI) tiles, reusable surface insulation blankets, thermal window panes, and thermal seals to protect the orbiter against aerodynamic heating.

The RSI tiles covering the orbiter are made of coated silica fiber (these tiles are over 96% pure silica) and the two types of RSI tiles differ only in surface coating to provide protection for different temperature regimes. The low-temperature (this is a relative concept) RSI consists of tiles, white in color, that cover portions of the vehicle where temperatures are less than 1,200°. This includes the majority of the upper surface of the fuselage and tail. The high-temperature RSI tiles are black in color, and cover portions of the orbiter where the temperatures reach up to 2,300°. There are two densities of high-temperature tiles available: one weighs 9 lbs. per cubic foot and is used in all high-temperature areas of the lower fuselage/wing and the sides of the forward fuselage except around the nose and main landing gear doors, all carbon-carbon interfaces and the ET umbilical doors. These areas use tiles weighing 22 lbs. per cubic foot. When the higher density tiles have been expended, new fibrous refractory insulation (FRCI) tiles will be used instead. The FRCI-12 tiles have a density of 12 lbs. per cubic foot and are more durable and resistant to cracking. Where temperatures reach above 2,300°, such as the wing leading edge and the nose cap, a composite of pyrolized carbon fibers in a pyrolized carbon matrix with a silicon carbide coating, commonly known as "carbon-carbon", is used. The use of carbon-carbon is being expanded, and current plans call for the high-temperature RSI tiles on the "chin" (lower forward fuselage) of all orbiters to be replaced by carbon-carbon plates during 1989. Where temperatures are less than 700°, and aerodynamic loads are minimal, such as the upper payload bay doors and lower aft fuselage, a flexible blanket consisting of coated Nomex felt is used. the use of these blankets is being expanded as flight experience is gained since they are easier to install and maintain than tiles. Each tile is unique to a location on the orbiter, being cut to size and shape to conform to the contours of the vehicle. The tiles are treated with a water-repellant for protection against moisture. As delivered, *Atlantis* was protected by 21,801 RSI tiles and 1,977 thermal protection blankets.

AVIONICS

Orbiter avionics consist of a fail-operational/fail-safe guidance, navigation and control system including three Singer-Kearfott KT-70/SKN-2600 type inertial measuring units; triple AIL manufactured Ku-band microwave scanning landing systems; three Northrop rate-gyro assemblies; three Hoffman L-band TACANS, three Bendix accelerometer assemblies; two Honeywell C-band radar altimeters; four AiResearch air data transducer assemblies; three Lear Siegler attitude direction indicators; two Collins horizontal situation displays; two Sperry Mach indicators; two Bendix barometric altimeters; and two Sperry 4096 Mode-C ATC transponders.

Communications and tracking equipment includes one or two Ku-band rendezvous radar/satellite communicators on the starboard side of the payload bay; two Ball star trackers on the forward fuselage; two Watkins-Johnson 100 Watt S-Band TWT amplifiers; two P-band UHF radios for EVA and ATC communications; two Conrac S-band FM transceivers for orbiter/ground and orbiter/payload; and a Ku-band transceiver for orbiter/ground communications.

Five identical IBM 4Pi/AP-101F General Purpose Computers (GPC) are interconnected through a modified Mil-STD-1553 digital data bus. During critical flight phases, four of the computers are assigned to perform guidance, navigation and control (GN&C) tasks acting as a cooperative, redundant set. Computations of each are verified by the others. In this way, the computer complex is able to achieve a very high system reliability, as well as a "fault tolerant" operational capability that lets the system endure component or connection failure in one or more of the computers without a loss of operational capability. The operational flight software was written by IBM Federal Systems Division. To guard against the possiblity of a software failure ("bug") affecting all four primary computers, the fifth computer is loaded with backup flight software (BFS) written by Rockwell International in Downey. This BFS is capable of guiding the shuttle safely back to earth, but can not support an operational mission. During orbital operations, the fifth computer usually is assigned to system management functions.

During non-critical flight periods in orbit, one computer is used for GN&C and control tasks, and another for system management. The remaining three can be used either for payload management, or deactivated as stand-by replacements.

Each GPC is equipped with 106,496 32bit words of main memory in a random-access, nonvolatile, destructive-readout ferrite core. Two 134 megabit mass memory tape units act as data and program storage devices. Four 5x7 in. monographic displays and keyboards provide the human interface to the computer system.

New upgraded GPCs, designated AP-101S, will replace the older models in late 1988 or early 1989. The upgraded computers will allow NASA to incorporate more capabilities into the orbiters, and apply advanced computer technologies that were not available when the orbiter first was designed. The new computer design began during January 1984, twelve years after the original computers were designed. The upgraded GPCs provide two-and-a-half times the existing memory capacity, and up to three times the existing processor speed. Minimal impact to flight software is expected. The new computers are half the size, weigh approximately half as much, and require less power to operate.

LANDING GEAR

The orbiter landing gear is arranged in a conventional tricycle configuration consisting of a nose landing gear, and a left and right main gear. When the nose landing gear is retracted, two doors cover the landing gear well. The main gear each have a single door that enclose the gear well.

The landing gear is constructed of high strength, stress/corrosion resistant steel and aluminum alloys, stainless steel, and aluminum-bronze. Urethane paint and cadmium-titanium plating is applied to all exposed steel surfaces, and conventional anodizing plus urethane paint is applied on exposed aluminum surfaces for protection during space flight.

The nose and main landing gear each contain a shock strut which is the primary source of shock attenuation during orbiter landing. The shock struts are conventional pneudraulic (gaseous nitrogen/hydraulic fluid) shock absorbers. A departure from convention is the separation of the gaseous nitrogen from the hydraulic fluid by a floating diaphragm to preclude the possibility of the nitrogen dispersing throughout the hydraulic fluid during zero-g operations. This separation is required to maintain proper shock strut performance at touchdown.

The nose gear retracts forward and up into the forward fuselage, while the main gear retracts forward and up into the wing. Each landing gear is held in the retracted position by an uplock hook. The uplock hooks are hydraulically released and the landing gear then free-falls, assisted by springs and hydraulic pressure. The applicable landing gear doors are opened through mechanical linkage attached to the landing gear. The landing gear will reach the fully extended position within a maximum of ten seconds, and is locked in the down position by spring loaded bungees. The nose gear door opening and gear extension also is assisted by a pyrotechnic actuator to assure gear deployment in the event of adverse air loads.

The nose and main landing gear strut actuators assist in the landing gear deployment through hydraulic pressure. The strut actuators include an oil snubber to control the rate of extension of the gear free fall, and thereby prevent damage to the landing gear and downlink linkage. The strut actuators also serve as the landing gear retract actuators.

The nose landing gear consists of two wheel and tire assemblies and is steerable. The nose gear tires are 32 x 8.8 with a rated static load of 23,700 lbs. and an inflation pressure of 300 psi. Nose gear tire life nominally is five landings. An improved nose wheel steering system was incorporated on *Columbia* for STS-32, and the system since has been fitted to *Discovery* and *Atlantis*. The modification allows a safe high-speed engagement of the nose wheel steering system, and provides positive lateral directional control of the orbiter during rollout in the presence of high crosswinds and blown tires.

Each of the main landing gear consists of two wheel and tire assemblies, and each wheel has a brake assembly with anti-skid protection. The main gear tires are 44.5x16 with a rated static load of 49,200 lbs. and an inflation pressure of 260 psi. Again, tire life nominally is five landings.

An on-going program to improve the landing gear of the orbiters has resulted in the incorporation of thicker main gear axles to provide a stiffer configuration that reduces brake-to-axle deflections and minimizes potential brake damage. Strain gauges also have been added to each nose and main wheel to monitor tire pressure before launch, on-orbit and during landing.

BRAKES

There are four Goodrich brake assemblies, one for each main gear wheel. Each assembly uses four beryllium rotors and three carbon lined stators. The stators are attached to a torque tube. The brakes were designed to absorb 36.5 million ft.-pounds of energy for normal stops, or one emergency stop of 55.5 million ft.-pounds. Each brake is fitted with a Hydro-Aire anti-skid system.

The brakes were designed for the original (1973) predicted weight of the orbiter, and since the "as-built" orbiters are somewhat heavier, the brakes have little or no margin. Three missions, Flights 5, 16 and 24 have experienced thermal stator damage. To counter this, the anti-skid circuitry has been modified to balance hydraulic pressure between adjacent brakes, instead of reducing pressure to the opposite brake if a flat tire is detected. Thicker, carbon-lined beryllium rotors have been incorporated, and a long-term program is in progress to replace the rotors with a carbon-carbon rotor to provide higher braking capacity by increasing maximum energy absorption.

EXTERNAL TANK

The External Tank (ET) contains the propellants for the Space Shuttle Main Engines, and is built by Martin Marietta Aerospace in Michoud, Louisiana. Each ET is 154.2 ft. long and 27.6 ft. in diameter.

The ET consists of the forward liquid oxygen (LO_2) tank, an unpressurized intertank, and a liquid hydrogen (LH_2) tank. The LO_2 tank is a 53.4 ft. long fusion-welded aluminum alloy structure designed to operate in a pressure range of 20 to 22 psig. It contains anti-slosh and anti-vortex baffles. An anti-geyser system was installed on the first five tanks only. The tank feeds a 17 in. feedline which conveys LO_2 through the intertank, then external to the ET to the aft ET/orbiter disconnect on the lower aft section of the orbiter. The LO_2 tank's wedge nose reduces aerodynamic drag and heating, and serves as a lightning rod for the shuttle assembly. Total capacity of the LO_2 tank is 19,182 cubic feet, which translates to 143,351 gals. or 1,361,936 lbs. of liquid oxygen. The empty weight of the LO_2 tank is 12,500 lbs.

The intertank is a 22.5 ft. long cylindrical aluminum alloy structure with flanges on each end for joining to the LO_2 and LH_2 tanks. The intertank houses the ET instrumentation components and provides an umbilical plate that interfaces with the ground facility for purge gas supply, hazardous gas detection and hydrogen gas boiloff during ground operations. The intertank is vented in flight. The SRB trust beam and fittings, which distribute the SRB loads to the LO_2 and LH_2 tanks are contained in the intertank.

The LH_2 tank is a 96.7 ft. long fusion-welded aluminum alloy structure designed to operate in a pressure range of 32 to 34 psia. The tank contains an anti-vortex baffle and siphon outlet to transmit LH_2 from the tank through a 17 in. feedline to the ET/orbiter disconnect. The LH_2 tank does not contain the complex anti-slosh baffles that are installed in the LO_2 tank for the simple reason that LH_2 is so light that sloshing does not induce significant forces. The LH_2 tank weighs 26,640 lbs. when empty, and can hold 385,000 gals. of LH_2 weighing 227,641 lbs. It has an internal volume of 51,737 cubic feet.

The ET is thermally protected with a nominal 1 in. spray-on foam insulation, and a charring ablator to withstand localized high heating during the boost phase. This insulation was painted white for the first two shuttle flights (STS-1 and STS-2), but has been left unpainted on subsequent flights to save 595 lbs. and $15,000. A "lightweight" ET entered production in time for *Challenger's* first flight (STS-6), weight being saved primarily by redesigning the anti-slosh baffling and eliminating the anti-geyser line used to circulate liquid oxygen during fill. The lightweight external tanks have an empty weight of 66,800 lbs., 10,300 lbs. lighter than the earlier tanks.

The ET is separated from the orbiter 18 seconds after main engine cut-off, just prior to reaching orbital velocity, and proceeds on a ballistic path for impact in the Indian Ocean. Gaseous oxygen is vented from the LO_2 vent valve in the nose of the ET to induce a tumbling motion (at a minimum of ten degrees per second end-over-end) that assures the ET breaks up in flight. The ET is the only major component of the space shuttle system that is not recovered and reused.

During September 1983 Martin-Marietta received a contract to build 26 ETs and provide long-lead materials for 21 more. This contract is valued at $505 million.

SOLID ROCKET BOOSTERS

Two Solid Rocket Boosters (SRB) burn for two minutes with the Space Shuttle Main Engines to provide initial ascent thrust. The motor casings are manufactured by McDonnell Douglas Astronautics Company in Huntington Beach, California, and the motor segments are produced by the Thiokol Chemical Corporation at Brigham City, Utah. The assembled SRB is 149 ft. long and 12 ft. 5 in. in diameter. Each SRB weighs approximately 1,286,000 lbs. and produces 3,060,000 lbs. of thrust at sea level (SRBs for STS-1 through STS-6 produced 2,940,000 lbs. thrust). The propellant grain is shaped to reduce thrust approximately 30% 55 seconds after lift-off to prevent overstressing the vehicle during the period of maximum aerodynamic pressure. The thrust vector control (TVC) system is hydraulically actuated, and has a maximum omni-axial gimbal capability of slightly over 7°, which, in conjunction with the SSME, provides flight control during the boost phase.

The Solid Rocket Booster is made up of several subassemblies: the nose cone, the Solid Rocket Motor, and the nozzle assembly. Each Solid Rocket Motor case is made of eleven individual cylindrical weldless D6AC steel sections. When assembled they form a tube almost 116 ft. long weighing 185,000 lbs. (empty). The 11 sections are the forward dome section, six cylindrical sections, the aft External Tank attachment ring section, two stiffener sections, and the aft dome section. The 11 sections are joined by tang-and clevis joints held together by 177 steel pins around the circumference of each joint. Segments used after STS-6 have walls that are 2 to 4 hundredths of an inch thinner than earlier segments, affording a weight reduction of 4,000 lbs. for an entire assembly. A case constructed of filament-wound composite, weighing 28,000 lbs. less than steel cases, was designed and built for operation from the Vandenberg Launch Site. Concerns over this casing's ability to withstand the bending forces generated after main engine ignition (while the SRBs are still bolted to the launch mount) have resulted in its not being adopted for use. The filament wound casings that were produced, such as those currently displayed with *Pathfinder* at MSFC, were not painted, and retained the natural black finish inherent to this construction technique.

After the sections have been machined to fine tolerances and fitted, they are partly assembled at the factory into four casting segments. These are the segments into which the propellant (TP-H1148) is poured (or cast). Joints assembled before the segment is filled with fuel are known as "factory" joints. Joints between the four casting segments are called "field" joints, and are assembled at the launch site. The case liner material is an asbestos filled carboxyl terminated polybutadiene (CPTB) polymer known as UF-2137.

Due to its part in the 51-L accident, the SRM field joint metal parts, internal case insulation and seals were redesigned, and a weather protection system added. In the 51-L design, the application of actuating pressure to the upstream face of the O-ring was essential for proper joint sealing performance because large sealing gaps were created by pressure-induced deflections, compounded by significantly reduced O-ring sealing performance at low temperature. The major change in the motor case is the new tang capture feature to provide a positive metal-to-metal interference fit around the circumference of the tang and clevis ends of the mating segments. The interference fit limits the deflection between the tang and clevis O-ring sealing surfaces caused by motor pressure and structural loads. The joints are designed so the seals will not leak under 200% design structural deflection and rates.

The new design, with the tang capture feature, controls the O-ring sealing gap dimensions. The sealing gap and the O-ring seals are designed so that a positive compression (squeeze) is always on the O-rings. The clevis O-ring groove dimension has been increased so that the O-ring never fills more than 90% of the groove. The new field joint design also includes an additional leak-check port to ensure that the primary O-ring is positioned properly. A new O-ring in the capture feature also serves as a thermal barrier in case the sealed insulation is breached. The field joint internal case insulation was modified to be sealed with a pressure-actuated flap called a J-seal, rather than putty as in the earlier configuration. Longer field joint case mating pins, with an improved retainer band, were added to improve the shear strength of the pins. External heaters with integral weather seals were incorporated to maintain the joint and O-ring temperature at a minimum 75° F. The weather seal also prevents water intrusion into the joint.

The TP-H1148 is a composite type solid propellant formulated of an aluminum powder (16%) as a fuel; ammonium perchlorate (69.93%) as an oxidizer; iron oxidizer powder (0.07%) as a burning rate catalyst; polybutadiene acrylic acid acrylonitrile (PBAN - 12.04%) as a rubber based binder; and an epoxy curing agent (1.96%). The binder and epoxy also are burned as fuel, adding a small amount of thrust.

The SRB is attached to the external tank at the forward end of the forward skirt by a single thrust attachment. The recovery parachute system is attached to the same thrust structure. A non-load bearing sway brace connects the aft portion of the SRB with the bottom of the external tank. While on the launch pad, four "hold-down" bolts secure each SRB to the mobile launch platform. These eight bolts are all that secure the space shuttle vehicle to the MLP. Each bolt is 28 in. long and 3.5 in. in diameter. The top nut of each bolt is frangible, containing two NASA Standard Initiators (NSIs) which are ignited at SRB ignition.

Each SRB has two integrated electronic assemblies, one forward and one aft. The forward assembly jettisons the nozzle after burnout, initiates the release of the nose cap and frustum, detaches the parachutes, and turns on the recovery aids. The aft assembly, mounted in the ET/SRB attach ring, connects with the forward assembly and the orbiter avionics system for SRB ignition commands and thrust vector control.

Rockwell International via Dennis Jenkins

"Atlantis" during roll-out at Rockwell International's Palmdale facility. Insulation blankets were used more extensively on this shuttle than any previous. Mock-up OMS pods were used in place of functional units.

The SRBs are released by pyrotechnic separation devices at the forward thrust attachment and the aft sway braces. Eight solid fuel separation rockets on each SRB (four aft and four forward) thrust for 1.02 seconds to provide positive separation of the SRB from the external tank. Each separation motor is 31.1 in. long and 12.8 in. in diameter. The SRBs reach a maximum altitude of 220,000 ft. before falling back towards the ocean. Following separation at 19,000 ft., the SRB nose cap is deployed for recovery initiation. A pilot chute deploys the drogue chute which, after stabilizing the SRB, deploys the main parachutes. The three main chutes inflate to a reefed condition at 8,800 ft., and are fully deployed by 3,400 ft. The SRBs impact the water approximately 160 miles downrange, where two recovery vessels insert plugs in the nozzle opening to prevent water from weighing them down, and tow them to port for refurbishment and reuse. Location aids are provided for each SRB, frustum/drogue chute, and main parachutes. These include a transmitter, antenna, strobe/converter, battery and salt water switch electronics. The location aids are designed for a minimum operating life of 72 hours and are considered usable up to 20 times by refurbishment. The strobe light is an exception, having an operating life of 280 hours. The battery is used only once.

RANGE SAFETY SYSTEM

Both Solid Rocket Boosters, and the External Tank are fitted with explosive charges capable of destroying the vehicle. These can be detonated on the command of a Range Safety Officer (RSO) if the vehicle crosses the limits established by flight analysis before launch, or if the vehicle is no longer in controlled flight. The determination of "controllability" is made by the Flight Director in Mission Control, Houston. An encoded "arm" command is sent to the range safety package, which is detonated by a subsequent encoded "fire" command. The system is considered completely fail-safe.

The range safety package consists of linear shaped charges on both the liquid oxygen and liquid hydrogen tanks on the ET, and linear shaped charges running the length of each SRB. The SRB charges split the SRB casing open, effectively terminating thrust.

LAUNCH FACILITIES

On April 14, 1972, Dr. James C. Fletcher announced the selection of the Kennedy Space Center and Vandenberg AFB as the sites from which space shuttle flights would be conducted.

KENNEDY SPACE CENTER (KSC)

The primary launch site for the space shuttle is Launch Complex 39 at the Kennedy Space Center on Florida's Atlantic coast. LC-39 was constructed to support the *Apollo* moon missions, and subsequently supported the *Skylab* missions and *Apollo-Soyuz* test project. A complete shuttle landing facility was built, and the existing launch facilities modified in the late 1970s to support the space shuttle program. Recommendations made after the *Challenger* accident indicate that most shuttle flights will continue to be recovered at Edwards AFB in California.

SHUTTLE LANDING FACILITY (SLF)

The SLF runway (15/33) is 15,000 ft. long, 300 ft. wide and 16 in. thick at the center. Safety overruns of 1,000 ft. are provided at each end. The runway has a 24 in. crown at the centerline, and small grooves, each 0.25 in. wide and deep have been cut in the concrete every 1.25 in. to facilitate rain removal and improve traction. A 550 by 490 ft. apron is attached to the southeastern end of the runway. The Mate/De-Mate Device, which lifts the orbiter for attachment to, or removal from, the Shuttle Carrier Aircraft, is located on the northeast corner of the ramp. TACAN and a microwave landing system (MLS) are provided.

VEHICLE ASSEMBLY BUILDING (VAB)

The VAB was built to assemble the *Saturn V* moon-rocket, and was at the time the largest building (enclosed space) in the world. It covers eight acres and has an enclosed volume of 129,428,000 cubic feet. It is 525 ft. high, 218 ft. long and 518 ft. wide. The structure is designed to withstand winds of up to 125 mph. The VAB's foundation rests on more than 4,200 steel pilings 16 in. in diameter, driven down to bedrock at a depth of 160 ft. The VAB has more than 70 lifting devices, including 250 ton bridge cranes used to lift the orbiter. The low bay area contains SSME maintenance and overhaul shops, office space and other shop areas. High Bays 1 and 3 are used for the integration and stacking of the complete space shuttle vehicle. High Bay 2 is used for ET checkout and storage, and as a contingency storage area for orbiters. High Bay 4 also is used for ET checkout and storage, as well as for vertical payload canister operations. During the *Apollo/Saturn* program this building was known as the *Vertical* Assembly Building, but still abbreviated VAB.

ORBITER PROCESSING FACILITY (OPF)

The OPF consists of two identical high bays connected by a low bay. The high bays are each 197 ft. long, 150 ft. wide and 95 ft. high. Each is equipped with two 30 ton bridge cranes and contain platforms which surround the orbiter to provide maintenance access. The low bay is 233 ft. long, 97 ft. wide and 25 ft. high, and houses electronic, mechanical and electrical support systems, as well as shops and office space. Most pre/post-flight maintenance and refurbishment is accomplished in the OPF. Payloads processed through checkout in a horizontal attitude, such as the Hubble Space Telescope, are loaded into the orbiter in the OPF. Payloads that are processed and installed vertically are mated with the orbiter at the launch pad.

ORBITER MODIFICATION AND REFURBISHMENT FACILITY (OMRF)

Located just northwest of the VAB, this 50,000 sq. ft. facility is used to perform modification, rehabilitation and overhaul of orbiters outside the facilities used for the normal operational flow. The building consists of a high bay identical in size to one of the OPF's, and a two-story low bay area.

LAUNCH CONTROL CENTER (LCC)

Four "Firing Rooms" are housed in the LCC, and are used to monitor and control "tests" during space shuttle turnaround and launch operations. Firing Room 1 and Control Room 3 provide services for launches, while Firing Room 2 and Control Room 4 serve as software

NASA

"Atlantis" following arrival at the Kennedy Space Center. It is seen being towed from the Orbiter Processing Facility to the VAB for mating with its external tank and SRBs. Functional OMS pods have replaced the mock-ups.

development sets, and provide control during OPF operations. The ground-based Launch processing System (LPS) consists of three major parts: the Central Data Subsystem (CDS), the Control, Checkout, and Monitor Subsystem (CCMS), and the Record and Playback Subsystem (RPS). The CDS consists of Honeywell mainframe computers that provide software development and off-line test support, as well as real-time recording during tests. The CCMS actually controls the launch of a shuttle, and is an extensively networked distributive processing system. Each firing room consists of about 50 minicomputers that monitor and control 50,000 measurements and commands during launch operations. The RPS records all raw telemetry for use in analysis and troubleshooting. The LCC was constructed for checking out and launching *Saturn* boosters.

OTHER FACILITIES

A new Logistics Facility provides 324,640 sq. ft. of storage for some 190,000 shuttle hardware parts. A state-of-the-art storage and retrieval system includes automated handling equipment to find and retrieve specific parts.

The Rotation Processing Building, located just north of the VAB, receives SRB segments shipped by rail from the manufacturer. Here, inspection, rotation and aft-skirt/aft-segment buildup are performed.

The two SURGE buildings each store two SRB flight sets (eight segments) after transfer from the adjacent Rotation Processing Building.

The SRB Assembly and Refurbishment Facility, located several miles south of the VAB, covers 45 acres and consists of four main buildings. It is here that inert components such as the aft and forward skirts, frustums, nose caps, recovery systems, etc. are received, refurbished, assembled and tested.

TRANSPORTABLE EQUIPMENT

The Mobile Launcher Platform (MLP) is 25 ft. high, 160 ft. long and 135 ft. wide, and serves as a transportable launch base for the space shuttle. The platform is constructed of steel up to six inches thick. Three openings through the main body of each platform allow the two SRBs and the SSMEs to exhaust through. A large Tail Service Mast (TSM) on each side of the orbiter aft fuselge provides umbilical connections to the spacecraft. LO_2 plumbing runs through one TSM, while LH_2 lines go through the other. Each TSM is 15 ft. long, 9 ft. wide and rises 31 ft. above the platform deck. Each of the three MLPs weighs 8,230,000 lbs. empty. These platforms were modified from their previous *Saturn* duties.

The two Crawler-Transporter tracked vehicles also were previously used during the *Saturn* program. The transporters are 20 ft. tall, 131 ft. long and 114 ft. wide and are used to carry the MLPs from the VAB to the launch pad and back. Their maximum speed unloaded is two miles per hour and while carrying the MLP and space shuttle they can reach one mile per hour. A crawler has eight tracks, each of which has 57 cleats. Each cleat weighs one ton. Unloaded, each transporter weighs 6,300,000 lbs. Power is provided by two 2,750 horsepower diesel engines. The engines drive four 1,000 kilowatt generators that provide electrical power to 16 traction motors (two per track).

The Crawler-Transporters move on a roadway 130 ft. wide consisting of two 40 ft. wide lanes separated by a 50 ft. wide median. The top surface is composed of 4 to 8 in. of crushed river gravel. The distance from the VAB to Pad 39A is about 3.5 miles, and to Pad 39B, 4.25 miles.

A Payload Canister provides restraint and environmental protection to the various payloads while in transit from payload processing facilities to either the launch pad (vertical payloads) or the OPF (horizontal). The canister is 69 ft. long, 21 ft. wide and 21 ft. high. It is constructed to physically resemble the inside of the orbiter's payload bay, and can accommodate any shuttle payload.

The Payload Canister Transporter is a 48 wheel self-propelled truck designed to support the canister and associated hardware. The transporter rides on rubber tires and is designed to operate on normal hard surfaced roads. It is 65 ft. long, 23 ft. wide and has an elevating flatbed that can be varied between 5.25 ft. and 7 ft. in height. Its wheels are independently steerable, and permit the transporter to move forward, backward or sideways; to "crab" diagonally; or to turn on its own axis. The transporter is driven by hydraulic motors powered by either a diesel engine (open-road) or an electric motor (indoors). The transporter weighs 140,000 lbs. empty. Top speed is 10 mph empty, or 5 mph with a full load. Because payload handling requires precise movements, the transporter has a "creep" mode that permits it to move as slowly as 0.014 mph.

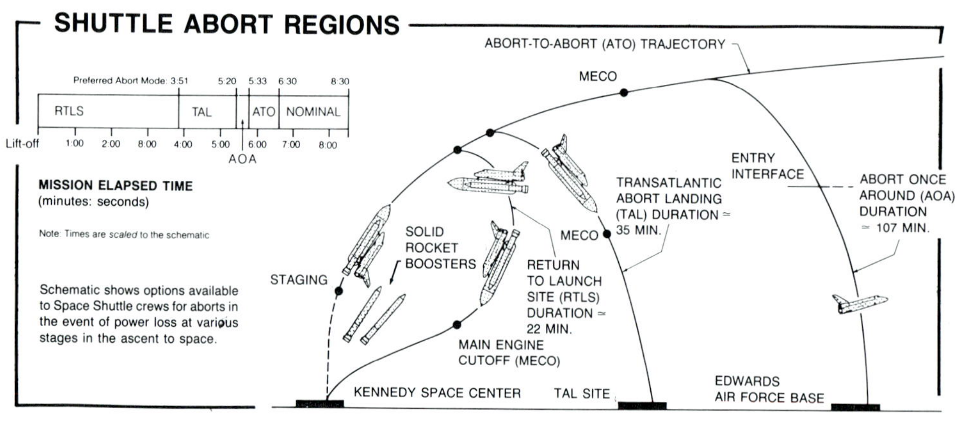

LAUNCH PADS 39A AND 39B

The Launch Complex 39 pads are roughly octagonal in shape, and each covers about 0.25 square miles. Shuttles are launched from the top of a concrete hardstand in the center of the pad. The Pad A stand is 48 ft. above sea level at its top, while Pad B is 55 ft. The top of the hardstand measures 390 by 325 ft. The two major items of equipment on each pad are the Fixed Service Structure (FSS) and the Rotating Service Structure (RSS).

The FSS is located on the west side of the hardstand. There are 12 work levels at 20 ft. intervals. The height to the top of the tower is 247 ft., and to the top of the lightning mast is 347 ft. The Orbiter Access Arm supports the "white room" that mates to the orbiter crew hatch. It is 65 ft. long, 5 ft. wide and 8 ft. high, and is attached to the FSS at the 147 ft. level. The FSS also supports the ET Gaseous Oxygen Vent Arm and the ET Hydrogen Vent Line Access Arm to allow the boiloff from the propellants in the ET to be vented away from the orbiter to prevent ice buildup. A white lightning mast rises 100 ft. above the Fixed Service Structure. A metal cable extends down to ground anchors 1,100 ft. south and north of the FSS.

The rotating service structure provides access to the orbiter for installation and servicing of payloads at the pad. It also supports the weather protection system that protects the orbiter from rain and hail while at the pad. The RSS pivots through one-third of a circle, from a retracted position well away from the shuttle to the point where its payload changeout room doors meet and match with the orbiter's payload bay doors. Most of its body is some 59 ft. above the pad. The rotating body is 102 ft. long, 50 ft. wide and 130 ft. high.

A flame trench 490 ft. long, 58 ft. wide and 40 ft. high divides the hardstand lengthwise from ground level to the pad surface. A one million pound flame deflector diverts the flames from the SRBs and SSMEs.

A slidewire system provides an escape route for the astronauts and closeout crew until the final 30 seconds of countdown. Five slidewires extend from the FSS at the Orbiter Access Arm level to hardened bunkers on the ground. A steel-wire basket 5 ft. in diameter, 42 in. deep and capable of holding two people is suspended from each wire. The 1,200 ft. slide to safety takes approximately 35 seconds.

A Sound Suppression Water System has been installed on the pad to protect the orbiter and its payload from damage by acoustical energy reflected from the MLP during launch. An elevated 300,000 gal. water tank located on the northeast side of the pad flows water through 7 ft. pipes for about 20 seconds after main engine ignition. The peak flow rate of 900,000 gals. per minute is reached at 9 seconds after lift-off. A Post-Shutdown Engine Deluge System provides 2,500 gals. per min. to cool the SSMEs in the event of an on-pad shutdown.

Propellant storage facilities are located at both pads. A 900,000 gal. LO2 dewar is located at the northwest corner, and transferred to the orbiter by two pumps at up to 10,000 gals. per minute. A 850,000 gal. LH2 dewar is located at the northeast corner, and transfers the fuel under pressure, LH2 being so light that pumps are not needed. Hypergolic propellants are stored in the southwest corner (MMH) and southeast corner (N2O4).

VANDENBERG LAUNCH SITE (VLS)

The dedicated Air Force launch site was to be Space Launch Complex Six (SLC-6) at Vandenberg AFB along California's central coast. The facility was originally constructed to support the Manned Orbiting Laboratory (MOL) program before its cancellation. After eight years and $3,800,000,000 of modifications, SLC-6 was ordered mothballed following the destruction of *Challenger*. The launch complex was declared "operational" during October 1985, but several conceptual and "as-built" flaws cast doubt on whether it could have supported a launch. Due to the amount of equipment removed from VLS to support other sites, it is doubtful if the facility could be reactivated.

MISSION CONTROL

Space shuttle control is conducted at two Mission Control rooms located in Building 30 at the Johnson Space Center in Houston, Texas. The responsibility for monitoring the orbiter is passed from the launch control center at KSC to mission control as soon as the vehicle reaches an altitude greater than the Fixed Service Structure ("clear of tower"). JSC then monitors the ascent, on-orbit and descent performance of the shuttle system.

MAJOR SPACE SHUTTLE STUDY CONTRACTS

CONTRACTOR (or team)	SCOPE OF STUDY	CUMULATIVE FUNDING (as of 31 March 1973)
Lockheed	Phase 'A' Vehicle Design and Development	2,500,000
General Dynamics	Phase 'A' Vehicle Design and Development	2,500,000
McDonnell Douglas	Phase 'A' Vehicle Design and Development	2,500,000
North American	Phase 'A' Vehicle Design and Development	2,500,000
Martin Marietta	Phase 'A' Vehicle Design and Development	2,500,000
McDonnell Douglas/Martin Marietta	Phase 'B' Vehicle Design and Development	19,300,000
North American/General Dynamics	Phase 'B' Vehicle Design and Development	19,300,000
Grumman Aerospace/Boeing Company	ASSC Vehicle Design and Development	13,600,000
Lockheed	ASSC Vehicle Design and Development	4,500,000
Chrysler	Recoverable Launch Booster	750,000
Chrysler	Pressure Fed Engine Booster	865,000
Mathematica	Economic Benefits	645,000
Aerojet General	High Pressure Engine (Orbiter)	6,700,000
Pratt & Whitney	High Pressure Engine (Orbiter)	6,700,000
Rocketdyne	High Pressure Engine (Orbiter)	6,700,000
Aerojet General	Pressure Fed Engine (Booster)	450,000
TRW Systems Group	Pressure Fed Engine (Booster)	450,000
United Technology	Solid Rocket Motor (Booster)	150,000
Aerojet	Solid Rocket Motor (Booster)	150,000
Lockheed	Solid Rocket Motor (Booster)	150,000
Thiokol	Solid Rocket Motor (Booster)	150,000
Aerospace Corporation	Operations and Payload Analysis	3,600,000
Lockheed	Payload Analysis	599,000
Parsons	KSC Faclities Planning	715,000

APPROACH AND LANDING TEST FLIGHTS (OV-101)

Captive-Inactive Flights:

Flight #	Date	OV Crew	Duration	Tailcone	Max Speed	Max Altitude
1	18-Feb-77	- - -	2 hrs 05 min	on	287	16,000
2	22-Feb-77	- - -	3 hrs 13 min	on	328	22,600
3	25-Feb-77	- - -	2 hrs 28 min	on	425	26,600
4	28-Feb-77	- - -	2 hrs 11 min	on	425	28,565
5	02-Mar-77	- - -	1 hr 39 min	on	272	30,000

Captive-Active Flights:

Flight #	Date	OV Crew	Duration	Tailcone	Max Speed	Max Altitude	Combined Weight
1	18-Jun-77	H/F	55 min 46 sec	on	208	14,970	508,000
2	28-Jun-77	E/T	62 min 00 sec	on	310	22,030	557,080
3	26-Jul-77	H/F	59 min 53 sec	on	311	30,292	565,600

Free-Flights:

Flight #	Date	OV Crew	Duration	Tailcone	Launch Speed	Launch Altitude	Landing Speed
1	12-Aug-77	H/F	5 min 21 sec	on	270	24,100	185
2	13-Sep-77	E/T	5 min 28 sec	on	269	26,000	194
3	23-Sep-77	H/F	5 min 34 sec	on	250	24,700	191
4	12-Oct-77	E/T	2 min 34 sec	off	240	22,400	199
5	26-Oct-77	H/F	2 min 01 sec	off	245	19,000	189

NOTES:

OV Crew H/F was Fred Haise and Gordon Fullerton.
OV Crew E/T was Joe Engle and Dick Truly.

SCA Crew for all flights included Fitzhugh Fulton, Jr (P), Thomas McMurtry (P), Louis Guidry, Jr. (FE) and Victor Horton (FE).

Speeds are in knots, altitudes in feet AGL, weights in pounds.

SIGNIFICANT ORBITER MILESTONES

MILESTONE	MPTA-098	*Challenger* STA-099	*Challenger* OV-099	*Enterprise* OV-101	*Columbia* OV-102	*Discovery* OV-103	*Atlantis* OV-104	OV-105
Start Structural assembly crew module *	—	28-Jan-77	21-Jun-79	04-Jun-74	28-Jun-76	03-Mar-80	15-Feb-82	•
Start Structural assembly aft-fuselage	24-Jun-75	14-Jun-76	—	26-Aug-74	13-Sep-76	10-Nov-80	23-Nov-81	•
Wings arrive at Palmdale from Grumman	—	16-Mar-77	—	23-May-75	26-Aug-77	30-Apr-82	13-Jun-83	Dec '86
Start final assembly	12-Jul-76	30-Sep-77	03-Nov-80	25-Aug-75	07-Nov-77	03-Sep-82	02-Dec-83	01-Aug-87
Complete final assembly	27-May-77	10-Feb-78	23-Oct-81	12-Mar-75	23-Apr-78	12-Aug-83	10-Apr-84	—
Rollout	—	14-Feb-78	30-Jun-82	17-Sep-76	08-Mar-79	16-Oct-83	06-Apr-85	—
Delivery to Kennedy Space Center	—	—	05-Jul-82	10-Apr-79	25-Mar-79	09-Nov-83	03-Apr-85	—
Flight Readiness Firing +	21-Apr-78	—	19-Dec-82	—	20-Feb-81	02-Jun-84	05-Sep-85	—
First Space Flight	—	—	04-Apr-83	—	12-Apr-81	30-Aug-84	03-Oct-85	early '92

* = simulated crew module on STA-99
\+ = first full-up static firing for MPTA-098
• = structural spares in storage at Downey
n/a = not applicable

SPACE SHUTTLE CONCEPTUAL DESIGN CONFIGURATION DATA

Phase:	A	A	A	A	A	B	B	B
Date:	01-Nov-69	11-Nov-69	22-Dec-69	31-Oct-69	22-Dec-69	11-Dec-69	11-Dec-69	13-Nov-70
Contractor:	MDC	MMC	Lockheed	GD	NAR	MDC/MMC	MDC/MMC	NAR/GD
Design:	- -	SpaceMaster		FR-3	- -	LCR	HCR	134C
ORBITER:								
Length Overall (ft):	107.00	160.00	163.00	179.00	202.00	147.60	158.30	186.50
Wing Span (ft):	75.70	107.00	92.00	146.00	146.00	113.80	97.50	106.25
Height (ft):	48.30	30.00	36.75	45.00	51.30	62.40	57.60	25.80
Wing Area (sq ft):	- -	1,812	- -	- -	2,830	1,375	- -	- -
Weight (empty - lbs):	- -	- -	183,830	287,000	256,515	188,600	203,500	215,672
# of Rocket Engines:	3	2	3	3	2	2	2	2
Thrust (Sea Level- lbs):	289,000	400,000	400,000	400,000	510,000	415,000	415,000	- -
# of Air-Breathers:	- -	0/4	4	2	4	4	4	2
Payload (pounds):	25,000	50,000	50,000	50,000	60,000	25,000	25,000	25,000
Payload Bay Size (ft):	15x30	15x60/22x30	15x60/22x30	15x60	15x60/22x60	- -	- -	- -
Cross-Range (nm):	- -	1,700	400 (1,500)	1,500	1,240	- -	- -	- -
External Tanks?:	none	none	none	none	none	none	none	none
BOOSTER:								
Length Overall (ft):	195.00	190.00	220.00	235.50	280.00	246.50	232.20	209.60
Wing Span (ft):	151.00	148.00	- -	- -	244.00	151.00	151.00	142.00
Height (ft):	69.50	58.00	- -	47.50	- -	56.90	56.90	50.00
Wing Area (sq ft):	- -	4,850	- -	- -	8,050	3,970	- -	- -
Weight (empty - lbs):	- -	448,369	322,550	517,000	587,522	452,800	483,900	- -
# of Rocket Engines:	15	14	13	15	11	13	14	12
Thrust (Sea Level -lbs):	289,000	400,000	400,000	400,000	510,000	415,000	415,000	- -
# of Air-Breathers:	- -	8	4	4	4	10	10	- -

Phase:		ASSC	ASSC	ASSC	Baseline	Award	Production
Date:	27-Apr-70	06-Jul-71	04-Jun-71	30-Jun-71	30-Aug-71	18-Feb-72	28-Apr-88
Contractor:	MSC/JSC	Grumman/Boeing	Lockheed	Chysler	MSC/JSC	MSC/JSC	RIC
Design:	DC-3	H33	LS-200	SERV	MSC-040	MSC-040C	OV-104
ORBITER:							
Length Overall (ft):	122.67	157.00	156.50	90.00 (dia)	121.88	123.75	122.17
Wing Span (ft):	90.80	97.00	92.00	n/a	75.00	80.20	78.06
Height (ft):	41.67	61.25	49.00	66.50	46.88	48.13	56.58
Wing Area (sq ft):	1,175	1,060	1,229	n/a	- -	- -	2,690
Weight (empty - lbs):	120,000	220,838	294,399	- -	- -	- -	151,205
# of Rocket Engines:	2	3	9	1	4	3	3
Thrust (Sea Level- lbs):	- -	415,000	530,000	5,800,000	325,000	375,000	375,000
# of Air-Breathers:	6	4	6	28	0	0	0
Payload (pounds):	15,000	60,000	65,000	88,060	60,000	60,000	60,000
Payload Bay Size (ft):	8x30	15x60	15x60	23x60	15x60	15x60	15x60
Cross-Range (nm):	200	1,100	1,500	n/a	1,500	1,500	1,085
External Tanks?:	none	LH2	LO2/LH2	none	LO2/LH2	LO2/LH2	LO2/LH2
BOOSTER:			none	none	SRB	SRB	SRB
Length Overall (ft):	203.00	245.00	n/a	n/a	- -	- -	149.16
Wing Span (ft):	141.00	177.50	n/a	n/a	n/a	n/a	n/a
Height (ft):	74.00	88.30	n/a	n/a	- -	- -	12.16 (dia)
Wing Area (sq ft):	2,840	- -	n/a	n/a	n/a	n/a	n/a
Weight (empty - lbs):	290,000	494,870	n/a	n/a	- -	- -	1,293,004
# of Rocket Engines:	11	12	n/a	n/a	2	2	2
Thrust (Sea Level - lbs):	- -	415,000	n/a	n/a	2,500,000	2,500,000	2,650,000
# of Air-Breathers:	4	8	n/a	n/a	- -	- -	0

-- = data not available
n/a = not applicable to this configuration

NASA

"Atlantis", fitted for STS-31, arriving at Pad 39A after the 3-1/2 mile journey from the VAB.

NASA

"Atlantis" during its STS-28 lift-off from the Kennedy Space Center on October 3, 1985.

NASA

"Discovery" lifts off on STS-26R, the first post-"Challenger" flight, heralding a new shuttle era.

AS-FLOWN STS MISSION DATA

Flight #	STS #	Manifest #	Orbiter Flight #	Dates Launch Landing	Times Launch Landing	Sites Launch Landing	Crew	Orbiter Launch Weight	Altitude Inclination	# of Orbits	Payload	Mission Duration	Notes
1	1		OV-102 1	12-Apr-81 14-Apr-81	07:04 10:21	KSC, Pad A Edwards AFB	John W. Young (C) Robert L. Crippen (P)	219,258	150 28.5	36	DFI	54 hrs 20 mins 57 secs 933,757 miles	Lift-off at 07:03:98.010 EST First orbital flight by a winged spacecraft 16 tiles lost, 148 damaged
2	2		OV-102 2	12-Nov-81 14-Nov-81	10:10 13:23	KSC, Pad A Edwards AFB	Joe H. Engle (C) Richard H. Truly (P)	230,708	164 28.5	36	OSTA-A MAPS SIR-A SMIRR FILE OCE NOSL DFI	54 hrs 24 mins 04 secs 933,757 miles	First re-use of a manned spacecraft First test of Remote Manipulator System Flight shortened due to Fuel Cell #1 failure 12 tiles damaged
3	3		OV-102 3	22-Mar-82 30-Mar-82	11:00 09:05	KSC, Pad A Northrup Strip	Jack R. Lousma (C) Charles Gordon Fullerton (P)	235,415	130 28.5	129	DFI	192 hrs 06 mins 09 secs 3,334,904 miles	Edwards landing site flooded by heavy rains 36 tiles lost, 19 damaged
4	4		OV-102 4	27-Jun-82 04-Jul-82	11:00 09:09	KSC, Pad A Edwards AFB	Thomas K. 'Ken' Mattingly II (C) Henry 'Hank' W. Hartsfield, Jr. (P)	241,664	160 28.5	112	DoD 82-1 CFES CIRRIS	169 hrs 10 mins 00 secs 2,900,000 miles	First DoD Mission SRBs not recovered due to chute malfunction First runway landing (Edwards runway 22)
5	5		OV-102 5	11-Nov-82 16-Nov-82	07:19 06:33	KSC, Pad A Edwards AFB	Vance D. Brand (C) Robert F. Overmeyer (P) Joseph P. Allen (MS) William B. Lenoir (MS)	246,997	160 28.5	81	ANIK-C3 SBS-C	122 hrs 14 mins 25 secs 1,800,000 miles	First operational flight Largest US crew flown to date
6	6		OV-099 1	04-Apr-83 09-Apr-83	13:30 10:53	KSC, Pad A Edwards AFB	Paul J. Weitz (C) Karol J. Bobko (P) F. Story Musgrave (MS) Donald H. Peterson (MS)	256,744	150 28.5	80	TDRSS-A CFES MLR/NOSL GAS (3)	120 hrs 24 mins 32 secs 2,530,567 miles	First Challenger flight First shuttle EVA First use of lightweight SRBs First use of lightweight ET
7	7		OV-099 2	18-Jun-83 24-Jun-83	07:33 06:57	KSC, Pad A Edwards AFB	Robert L. Crippen (C) Frederick H. Hauck (P) John M. Fabian (MS) Sally K. Ride (MS) Norman E. Thadard (MS)	249,178	150 28.5	97	ANIK-C2 PALAPA-B1 SPAS-01 OSTA-2 MLR CFES GAS (7)	146 hrs 24 mins 10 secs 2,530,567 miles	First US woman in space First shuttle rendevous
8	8		OV-099 3	30-Aug-83 05-Sep-83	02:30 12:40	KSC, Pad A Edwards AFB	Richard H. Truly (C) Daniel C. Brandenstein (P) Guion S. Bluford, Jr. (MS) Dale A. Gardner (MS) William E. Thorton (MS)	242,742	160 28.5	97	INSAT-1B PDRS/PFTA CFES OIM MLR GAS (7)	145 hrs 08 mins 40 secs 2,514,478 miles	First night launch and landing
9	9	(41-A)	OV-102 6	28-Nov-83 08-Dec-83	11:00 15:45	KSC, Pad A Edwards AFB	John W. Young (C) Brewster H. Shaw, Jr. (P) Robert A. Parker (MS) Owen K. Garriot (MS) Bryon K. Lichtenberg (PS) Ulf Merbold (PS)	247,619	135 57.0	167	Spacelab-1	247 hrs 47 mins 24 secs 4,295,853 miles	First Spacelab mission Flight delayed one month by SRB problem
10	11	41-B	OV-099 4	03-Feb-84 11-Feb-84	08:00 07:17	KSC, Pad A KSC, FL	Vance D. Brand (C) Robert L. 'Hoot' Gibson (P) Bruce McCandless II (MS) Ronald E. McNair (MS) Robert L. Stewart (MS)	250,452	165 28.5	127	PALAPA-B2 WESTAR-6 ACES IEF RME MLR SSIP (1) IRT GAS (5)	191 hrs 17 mins 54 secs 3,311,380 miles	First MMU test First KSC landing
11	13	41-C	OV-099 5	06-Apr-84 13-Apr-84	08:58 05:38	KSC, Pad A Edwards AFB	Robert L. Crippen (C) Francis R. 'Dick' Scobee (P) Terry J. Hart (MS) James D. van Hoften (MS) George D. Nelson (MS)	254,254	288 28.5	108	LDEF-1 SSIP (1) RME IMAX Camera	167 hrs 40 mins 07 secs 2,870,000 miles	First satellite repair (Solar Maximum) First operational use of MMUs First 'Direct Ascent Trajectory'
12	16	41-D	OV-103 1	30-Aug-84 05-Sep-84	08:47 06:37	KSC, Pad A Edwards AFB	Henry 'Hank' W. Hartsfield, Jr. (C) Michael L. Coats (P) Richard M. Mullane (MS) Steven A. Hawley (MS) Judith A. Resnik (MS) Charles D. Walker (MS)	216,778	160 28.5	97	SBS-D TELSTAR-3C Leasat-1 OAST-1 CFES RME SSIP (1) CLOADS IMAX Camera	144 hrs 56 mins 04 secs 2,210,000 miles	First 3-satellite deployment First Discovery flight
13	17	41-G	OV-099 6	05-Oct-84 13-Oct-84	07:03 12:26	KSC, Pad A KSC, FL	Robert L. Crippen (C) Jon A. McBride (P) Sally K. Ride (MS) Kathryn D. Sullivan (MS) David C. Leestma (MS) Marc Garneau (PS) Paul D. Scully-Power (PS)	241,780	190 57.0	133	OSTA-3 ERBS LFC/ORS RME TLD APE CANEX IMAX Camera	197 hrs 23 mins 32 secs 3,434,684 miles	EVA refueling simulation
14	19	51-A	OV-103 2	08-Nov-84 16-Nov-84	07:15 07:00	KSC, Pad A KSC, FL	Frederick H. Hauck (C) David M. Walker (P) Joseph P. Allen (MS) Anna L. Fisher (MS) Dale A. Gardner (MS)	261,697	160 28.5	127	ANIK-D2 Leasat-2 DMOS RME	191 hrs 45 mins 54 secs 3,289,406 miles	First satellite retrieval (WESTAR-6 and PALAPA-B2)
15	20	51-C	OV-103 3	24-Jan-85 27-Jan-85	14:40 16:23	KSC, Pad A KSC, FL	Thomas K. 'Ken' Mattingly II (C) Loren J. Shriver (P) Ellison S. Onizuka (MS) James F. Buchli (MS) Gary E. Payton (PS)	250,891	220	47	DoD	73 hrs 33 mins 13 sec	Classified DoD mission
16	23	51-D	OV-103 4	12-Apr-85 19-Apr-85	08:59 08:55	KSC, Pad A KSC, FL	Karol J. Bobko (C) Donald E. Williams (P) Margaret Rhea Seddon (MS) S. David Griggs (MS) Jeffrey A. Hoffman (MS) Charles D. Walker (PS) Senator E. J. 'Jake' Garn (PS)	248,927	250 28.5	110	Leasat-3 ANIK-E2 CFES AFE PPE/SAS SSIP (2) GAS (2)	167 hrs 55 mins 31 secs 2,889,785 miles	Mission extended two days
17	24	51-B	OV-099 7	29-Apr-85 06-May-85	12:02 12:11	KSC, Pad A Edwards AFB	Robert F. Overmeyer (C) Frederick D. Gregory (P) Don Leslie Lind (MS) William E. Thorton (MS) Norman E. Thagard (MS) Taylor G. Wang (PS) Lodewijk van den Berg (PS)	246,880	219 57.0	111	Spacelab-3	168 hrs 08 mins 46 secs 2,890,383 miles	First crosswind landing
18	25	51-G	OV-103 5	17-Jun-85 24-Jun-85	07:33 09:11	KSC, Pad A Edwards AFB	Daniel C. Brandenstein (C) John O. Creighton (P) Steven R. Nagel (MS) John M. Fabian (MS) Shannon W. Lucid (MS) Salman Abdul aziz Al-Saud (PS) Patrick Baudry (PS)	256,524	190 28.5	112	MORELOS-A ARABSAT-1B TELSTAR-3D Spartan-1 FEE FPE ADSF GAS (6)	169 hrs 38 mins 53 secs 2,916,127 miles	
19	26	51-F	OV-099 8	29-Jul-85 06-Aug-85	17:00 15:45	KSC, Pad A Edwards AFB	Charles Gordon Fullerton (C) Roy D. Bridges, Jr. (P) Karl G. Henize (MS) Anthony W. England (MS) F. Story Musgrave (MS) Loren W. Acton (PS) John-David F. Bartoe (PS)	252,855	207 49.5	126	Spacelab-2 SAREX CBDE PGU	190 hrs 45 mins 26 secs 3,283,543 miles	In-flight abort-to-orbit successful after one SSME shutdown
20	27	51-I	OV-103 6	27-Aug-85 03-Sep-85	06:58 09:16	KSC, Pad A Edwards AFB	Joe H. Engle (C) Richard O. Covey (P) James D. van Hoften (MS) William F. Fisher (MS) John M. Lounge (MS)		276 28.5	112	ASC-1 AUSSAT-1 Leasat-4 PVTOS	180 hrs 18 mins 29 secs 2,919,576 miles	In orbit repair of Leasat-2
21	28	51-J	OV-104 1	03-Oct-85 07-Oct-85	11:15 13:00	KSC, Pad A Edwards AFB	Karol J. Bobko (C) Ronald J. Grabe (P) Daivd C. Hilmers (MS) Robert L. Stewart (MS) William A. Pailes (PS)		320	111	DoD	97 hrs 45 mins 38 secs	First Atlantis flight
22	30	61-A	OV-099 9	30-Oct-85 06-Nov-85	12:00 12:45	KSC, Pad A Edwards AFB	Henry 'Hank' W. Hartsfield, Jr. (C) Steven R. Nagel (P) Bonnie J. Dunbar (MS) James F. Buchli (MS) Guion S. Bluford, Jr. (MS) Ernst Messerschmid (PS) Reinhard Furrer (PS) Wubbo Ockels (PS)	213,070	201 57.0	111	Spacelab-D1 GLOMAR	168 hrs 44 mins 51 secs 2,528,113 miles	Largest flight crew in history
23	31	61-B	OV-104 2	26-Nov-85 03-Dec-85	19:29 16:33	KSC, Pad A Edwards AFB	Brewster H. Shaw (C) Byran D. O'Conner (P) Sherwood C. Spring (MS) Mary L. Cleve (MS) Jerry L. Ross (MS) Charles D. Walker (PS) Rudolfo Neri Vela (PS)	261,455	190 28.45	109	MORELOS-B SATCOM-Ku1 AUSSAT-2 EASE/ACCESS CFES UVX IMAX Camera GAS	165 hrs 04 mins 50 secs 2,838,972 miles	First construction of structures in orbit Second night launch
24	32	61-C	OV-102 7	12-Jan-86 18-Jan-86	06:55 08:59	KSC, Pad A Edwards AFB	Robert L. 'Hoot' Gibson (C) Charles F. Bolden, Jr. (P) George D. Nelson (MS) Steven A. Hawley (MS) Franklin R. Chang-Diaz (PS) Robert J. Cenker (PS) Congressman C. William Nelson (PS)	255,471	160 28.45	96	SATCOM-Ku2 Leasat-5 MSL-2 CHAMP IR-IE SSIP (3) GAS (13)	146 hrs 04 mins 09 secs 2,528,658 miles	First flight of 102 following modifications First flight of SEADS, SILTS and SUMS flight test package
25	33	51-L	OV-099 10	28-Jan-86	11:38	KSC, Pad B —	Francis R. 'Dick' Scobee (C) Michael J. Smith (P) Ellison S. Onizuka (MS) Judith A Resnik (MS) Ronald E. McNair (MS) Gregory Jarvis (MS) Sharon Christa McAuliffe (SFP)	268,471			TDRSS-B Sparten/Halley MPESS CHAMP FDE RME TISP PPE SSIP (3)	1 min 13 secs	Vehicle exploded 73 secons after launch due to O-ring failure in right SRB. Crew, vehicle and payload lost.

ORBITER DATA

Orbiter	Empty Weight (w/o SSME)	Rollout	Delivery to KSC	First Flight
099 - *Challenger*	155,400	30-Jun-82	05-Jul-82	04-Apr-83
101 - *Enterprise*	150,000	17-Sep-76	10-Apr-79	—
102 - *Columbia*	157,289	08-Mar-79	25-Mar-79	12-Apr-81
103 - *Discovery*	151,419	16-Oct-83	09-Nov-83	30-Aug-84
104 - *Atlantis*	151,205	06-Apr-85	03-Ap.-85	03-Oct-85

NOTE: Enterprise Empty Weight includes ballast.

MMU FLIGHT HISTORY

Flight/Mission	EVA	Pilot	MMU#	Flight Time
10/41-B	1	McCandless	3	1 hrs 22 mins
	1	Stewart	2	1 hrs 09 mins
	2	McCandless	3	0 hrs 47 mins
	2	Stewart	2	0 hrs 44 mins
	3	McCandless	3	1 hrs 08 mins
		41-B Total:		5 hrs 10 mins
11/41-C	1	Nelson	2	0 hrs 42 mins
	2	van Hoften	3	0 hrs 28 mins
		41-C Total:		1 hrs 10 mins
14/51-A	1	Allen	3	2 hrs 22 mins
	2	Gardner	2	1 hrs 40 mins
		51-A Total:		4 hrs 02 mins

Total MMU #2 Flight Time:	3 hrs 53 mins
Total MMU #3 Flight Time:	6 hrs 29 mins
Total MMU Flight Time:	10 hrs 22 mins

MMU DATA

Size: 49.36 inches high, 32.56 inches wide, 47.6 inches deep with hand controllers extended.

Weight: 338 pounds loaded with 26 pounds of nitrogen propellant.

Prop. Tanks: Two, each holding 13 pounds of nitrogen pressurized at 3,000 pounds per square inch.

Thrusters: Total of 24. Each set of 12 powered by serarate electrical and propulsion systems.

Thrust: Each thruster delivers 1.7 pounds of thrust.

Speed: Operational: 1/3 to 1 mph. Top speed: 40 mph.

MAJOR COUNTDOWN MILESTONES

Time:		Milestone:
T-73:00:00.000		Call to Stations
T-61:00:00.000		Pressurize OMS and RCS tanks
T-58:00:00.000		Purge fuel cells with gaseous hydrogen and oxygen reactants
T-50:00:00.000		Mass Memory Unit (MMU) patch and compare
T-35:00:00.000		8 hour built-in hold
T-32:00:00.000		Load fuel cell liquid hydrogen and liquid oxygen
T-27:00:00.000		8 hour built-in hold
T-20:00:00.000		Retract rotating service structure
T-19:00:00.000		Interface test with Merritt Island (MILA) tracking station and Houston
T-10:00:00.000		Fill sound suppression water tank and load film cameras
T-08:00:00.000		12 hour, 10 minute built-in hold
T-07:00:00.000		Clear pad of non-essential personnel
T-05:00:00.000		Start terminal count. Start cyrogenic line chilldown in preparation for ET loading
T-04:30:00.000		Start filling LO2 tank
T-03:30:00.000		Wake flight crew for breakfast
T-02:50:00.000		Start filling LH2 tank
T-02:15:00.000		External tank filled, start replenish
T-02:05:00.000		1 hour built in-hold
T-01:50:00.000		Crew enters vehicle
T-01:10:00.000		Crew hatch closed
T-00:51:00.000		Inertial measurement unit pre-flight alignment
T-00:30:00.000		Start OMS pressurization
T-00:22:00.000		GPC patch and compare
T-00:20:00.000		10 minute built-in hold
T-00:20:00.000		GPCs transition to launch configuration (after hold terminates)
T-00:16:00.000		Start MPS helium pressurization
T-00:09:00.000		10 minute built-in hold
T-00:09:00.000		Start ground launch sequencer (after hold teminates)
T-00:07:00.000		Retract orbiter access arm
T-00:05:00.000		Start orbiter APUs and arm SRB ignition and RSS systems
T-00:03:30.000		Transfer orbiter to internal power
T-00:03:00.000		SSME gimbal check
T-00:02:55.000		Start pressurizing LO2 tank, retract gaseous oxygen vent arm
T-00:02:35.000		Transfer to on-board fuel cell reactants
T-00:01:57.000		Start pressurizing LH2 tank
T-00:00:28.000		Transfer control from Ground Launch Sequencer (GLS) to on-board Redundant Set Launch Sequencer (RSLS)
T-00:00:25.000		Start SRB APUs
T-00:00:06.566		SSME #1 ignition commanded
T-00:00:06.446		SSME #2 ignition commanded
T-00:00:06.326		SSME #3 ignition commanded
T-00:00:03.000		All three SSMEs at 90% power
T-00:00:00.000		SRB ignition (this is defined as T-zero)
T+00:00:00.008		Holddown post release
T+00:00:00.250		**Lift-off (first continuous vertical motion)**
T+00:00:06.500		Clear of launch tower (control passes to JSC)
T+00:00:07.300		Pitchover
T+00:00:59.000	+	Maximum dynamic pressure (max q)
T+00:02:06.000	+	SRB seratation (nominal) 163,000 feet altitude, 25 NM downrange
T+00:04:20.000	+	Return-to-Launch-Site (RTLS) abort no longer possible
T+00:05:50.000	+	One-engine, press to MECO (can abort to orbit if 1 SSME fails)
T+00:07:00.000	+	Two-engines, press to MECO (can abort to orbit if 2 SSMEs fail)
T+00:08:34.000	+	MECO (main engine cutoff) 57 NM altitude, 761 NM downrange (nominal)
T+00:08:52.000	+	ET separation
T+00:10:34.000	+	OMS-1 burn to normallize orbit
T+00:40:51.000	+	OMS-2 burn

+ = time dependant on flight profile

IN DETAIL:

Rockwell via Erik Simonsen

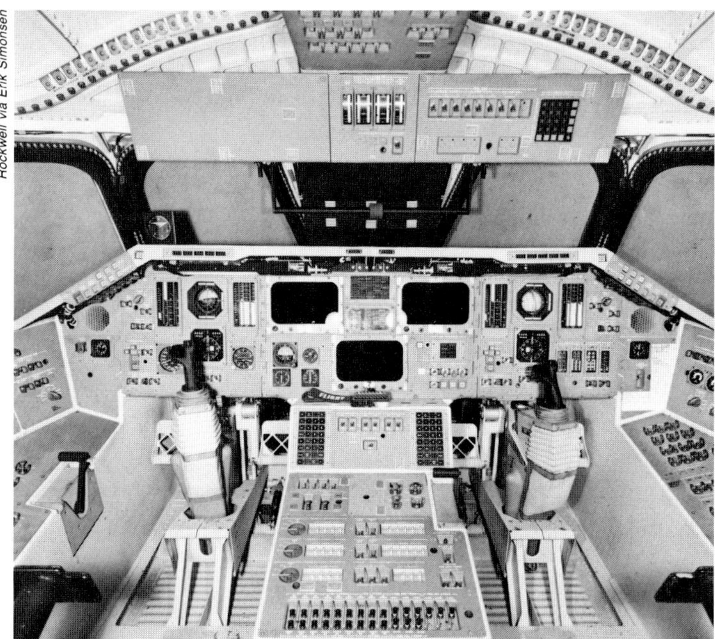

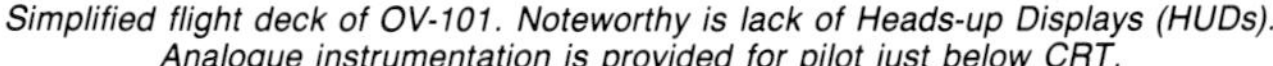

Simplified flight deck of OV-101. Noteworthy is lack of Heads-up Displays (HUDs). Analogue instrumentation is provided for pilot just below CRT.

Rockwell via Erik Simonsen

OV-101 was equipped with Lockheed zero-zero ejection seats during atmospheric flight test phase. Seats also were used by "Columbia" during early orbital test flights.

Rockwell via Gerald Balzer

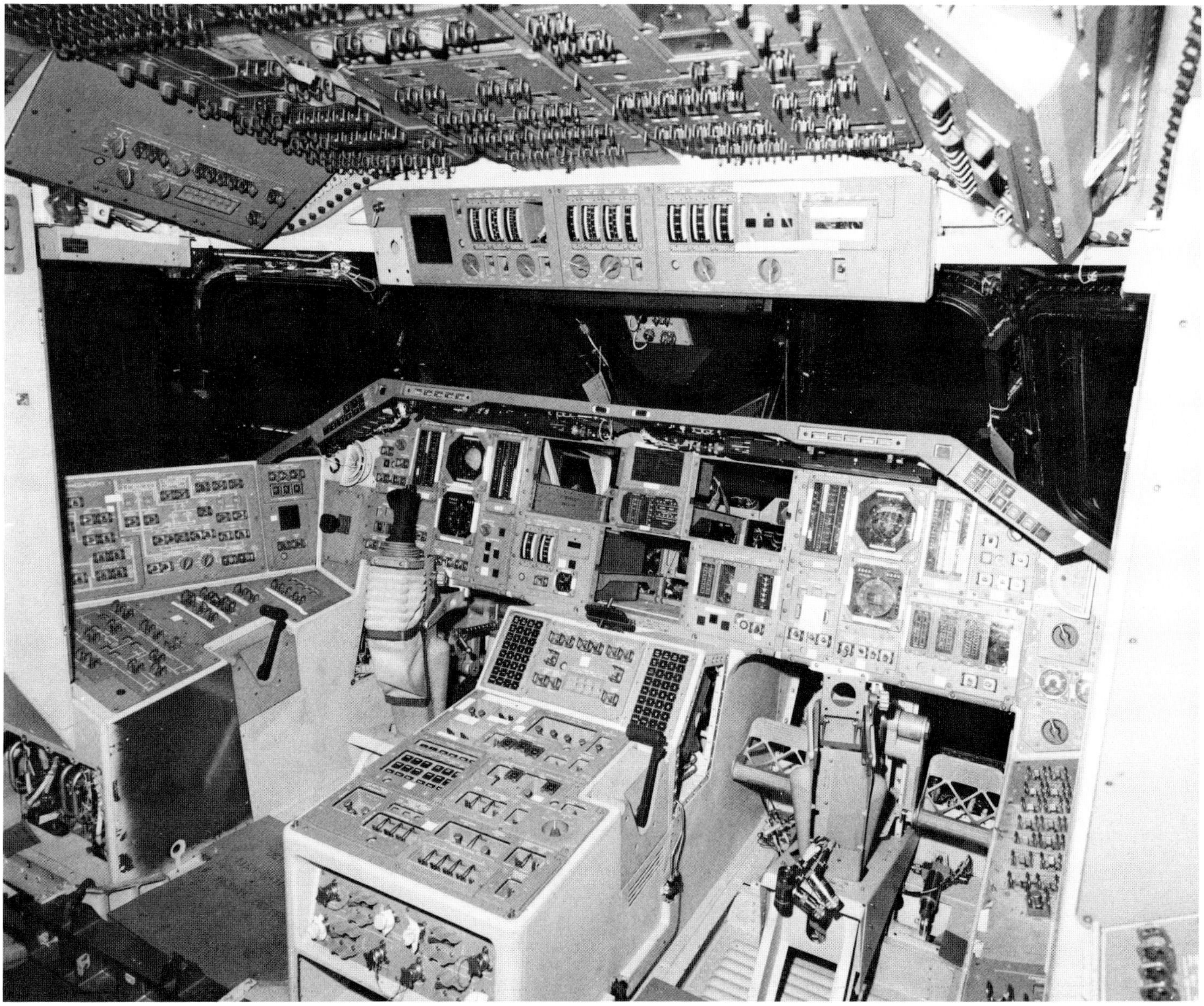

"Columbia" showing the extensive overhead switch collection. The three CRTs are missing from the center of the instrument panel. Lack of HUDs is noteworthy. All switches are either recessed or protected by partial covers to prevent accidental tripping by crew members floating in zero-gravity.

Rockwell via Erik Simonsen

"Challenger" flight deck is representative of production orbiters. HUD locations are clearly visible. Switch next to each key pad determines CRT the keypad is switched to. Either key pad may be used to communicate with middle CRT. Switches above and left of the circular switch on the lower center console command manual ET and SRB separation.

Rockwell via Erik Simonsen

"Challenger" showing one of the HUDs in place. Checklists and procedures pasted on every available surface around the astronaut are readily visible. The three covered switches in the extreme lower left are used to manually shut down the SSMEs in an emergency.

NASA via Gerald Balzer

Astronaut John M. Fabian in the aft flight deck area of "Columbia" during STS-5. Control stick is used to command the RMS. Rear facing windows look into the payload bay.

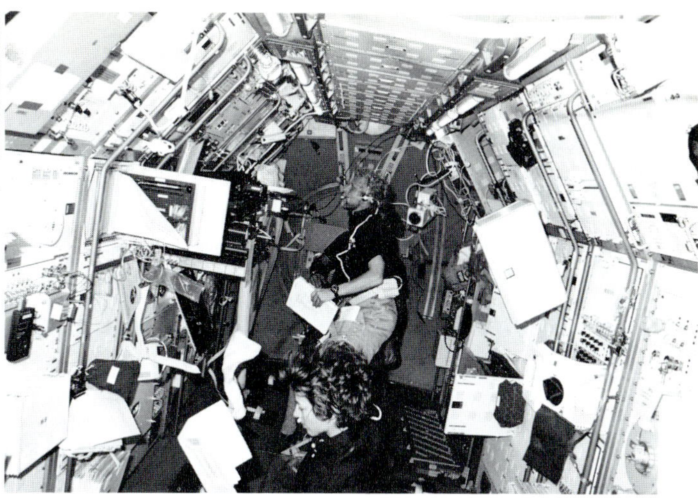
NASA via Gerald Balzer

STS-30 (61-A) carried the largest flight crew in history. Astronauts Bonnie Dunbar and Reinhard Furrer are seen in fully pressurized Spacelab D-1 in the shuttle payload bay.

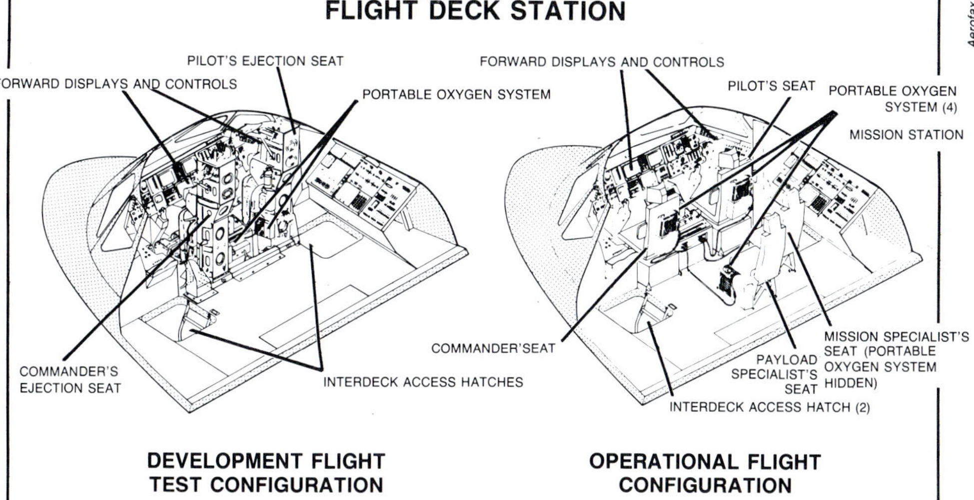

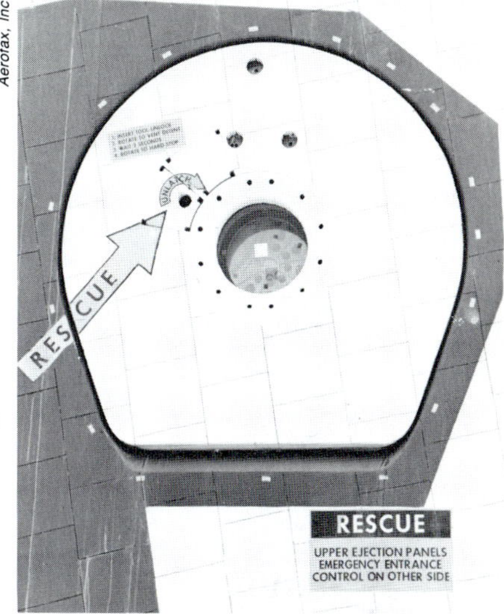

Aerofax, Inc.

The crew hatch of OV-101, "Enterprise". The 40-inch hatch is hinged at the bottom and opens outward.

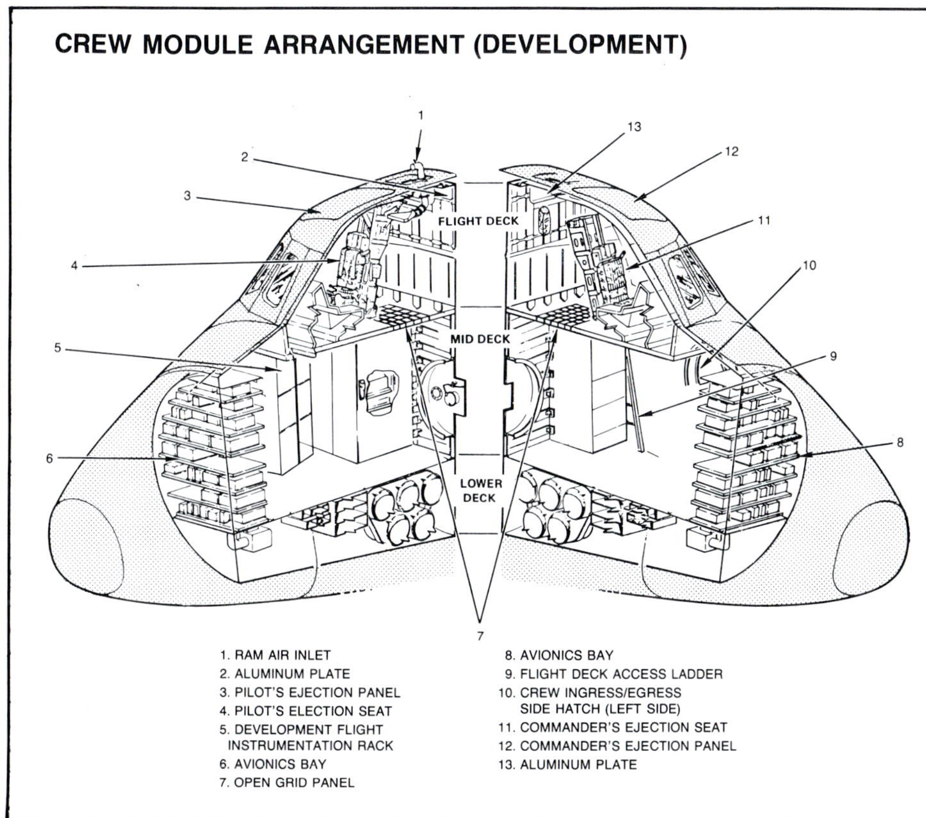

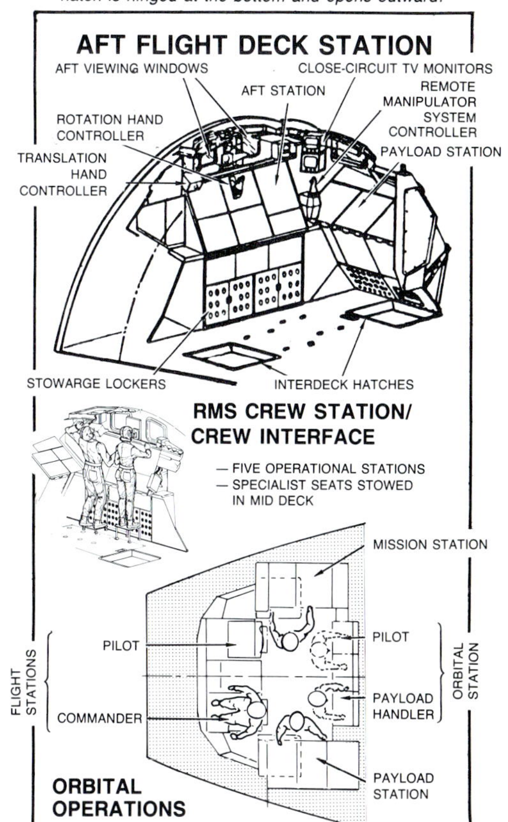

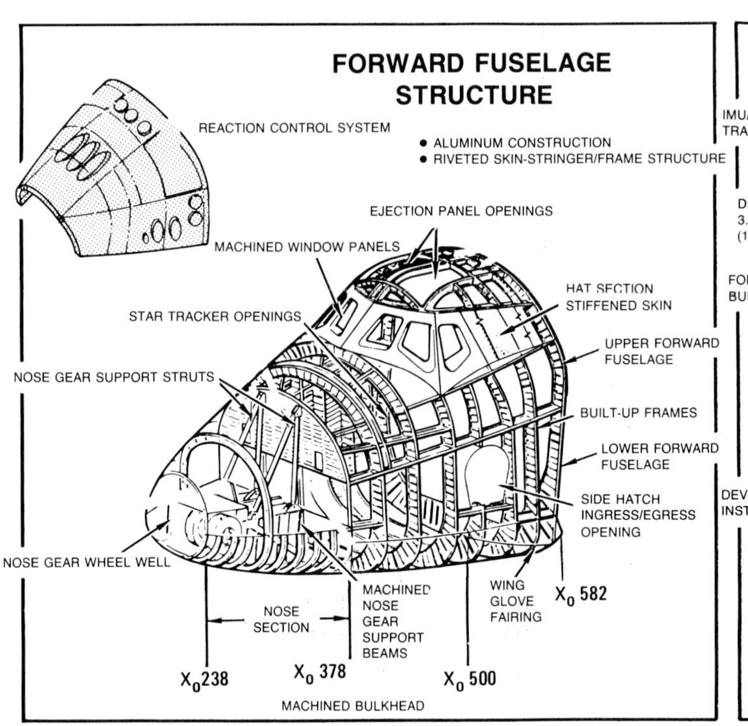
FORWARD FUSELAGE STRUCTURE
REACTION CONTROL SYSTEM
• ALUMINUM CONSTRUCTION
• RIVETED SKIN-STRINGER/FRAME STRUCTURE
EJECTION PANEL OPENINGS
MACHINED WINDOW PANELS
HAT SECTION STIFFENED SKIN
STAR TRACKER OPENINGS
UPPER FORWARD FUSELAGE
NOSE GEAR SUPPORT STRUTS
BUILT-UP FRAMES
LOWER FORWARD FUSELAGE
SIDE HATCH INGRESS/EGRESS OPENING
NOSE GEAR WHEEL WELL
MACHINED NOSE GEAR SUPPORT BEAMS
WING GLOVE FAIRING
X_0 582
NOSE SECTION
$X_0$238
X_0 378
X_0 500
MACHINED BULKHEAD

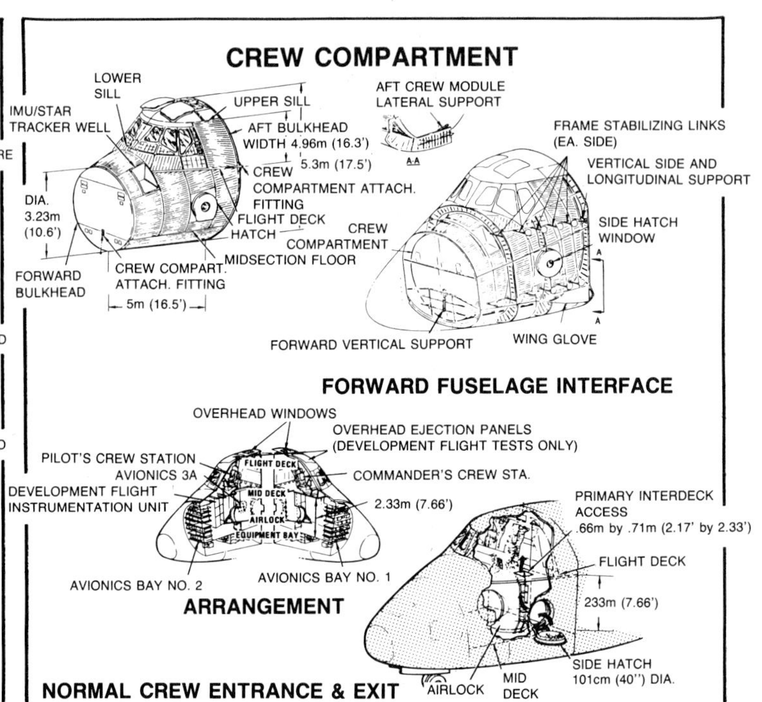
CREW COMPARTMENT
LOWER SILL
IMU/STAR TRACKER WELL
UPPER SILL
AFT CREW MODULE LATERAL SUPPORT
AFT BULKHEAD WIDTH 4.96m (16.3')
5.3m (17.5')
CREW COMPARTMENT ATTACH. FITTING
FRAME STABILIZING LINKS (EA. SIDE)
VERTICAL SIDE AND LONGITUDINAL SUPPORT
DIA. 3.23m (10.6')
FLIGHT DECK HATCH
CREW COMPARTMENT MIDSECTION FLOOR
SIDE HATCH WINDOW
FORWARD BULKHEAD
CREW COMPART. ATTACH. FITTING
5m (16.5')
FORWARD VERTICAL SUPPORT
WING GLOVE
FORWARD FUSELAGE INTERFACE
OVERHEAD WINDOWS
OVERHEAD EJECTION PANELS (DEVELOPMENT FLIGHT TESTS ONLY)
PILOT'S CREW STATION
AVIONICS 3A
COMMANDER'S CREW STA.
DEVELOPMENT FLIGHT INSTRUMENTATION UNIT
FLIGHT DECK
MID DECK
AIRLOCK
EQUIPMENT BAY
2.33m (7.66')
PRIMARY INTERDECK ACCESS .66m by .71m (2.17' by 2.33')
FLIGHT DECK
AVIONICS BAY NO. 2
AVIONICS BAY NO. 1
ARRANGEMENT
233m (7.66')
SIDE HATCH 101cm (40") DIA.
NORMAL CREW ENTRANCE & EXIT
AIRLOCK
MID DECK

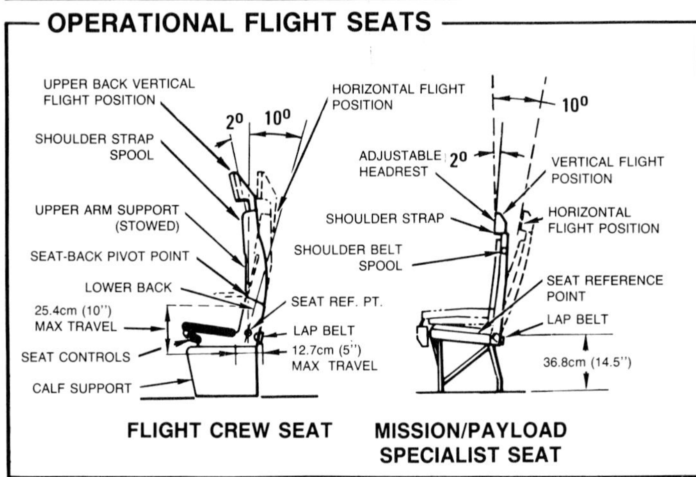
OPERATIONAL FLIGHT SEATS
UPPER BACK VERTICAL FLIGHT POSITION
2°
10°
HORIZONTAL FLIGHT POSITION
SHOULDER STRAP SPOOL
UPPER ARM SUPPORT (STOWED)
SEAT-BACK PIVOT POINT
LOWER BACK
25.4cm (10") MAX TRAVEL
SEAT REF. PT.
LAP BELT
SEAT CONTROLS
12.7cm (5") MAX TRAVEL
CALF SUPPORT
ADJUSTABLE HEADREST
VERTICAL FLIGHT POSITION
SHOULDER STRAP
HORIZONTAL FLIGHT POSITION
SHOULDER BELT SPOOL
SEAT REFERENCE POINT
LAP BELT
36.8cm (14.5")
FLIGHT CREW SEAT
MISSION/PAYLOAD SPECIALIST SEAT

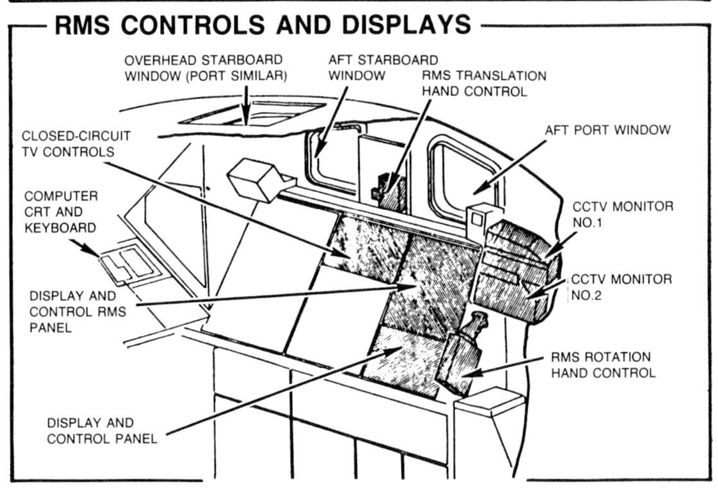
RMS CONTROLS AND DISPLAYS
OVERHEAD STARBOARD WINDOW (PORT SIMILAR)
AFT STARBOARD WINDOW
RMS TRANSLATION HAND CONTROL
AFT PORT WINDOW
CLOSED-CIRCUIT TV CONTROLS
COMPUTER CRT AND KEYBOARD
CCTV MONITOR NO.1
CCTV MONITOR NO.2
DISPLAY AND CONTROL RMS PANEL
RMS ROTATION HAND CONTROL
DISPLAY AND CONTROL PANEL

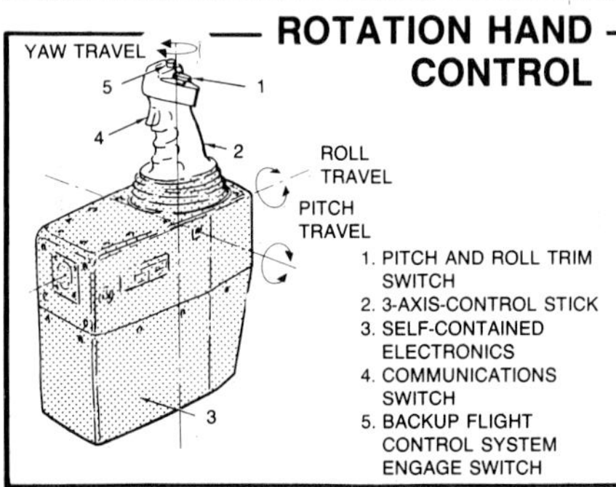
ROTATION HAND CONTROL
YAW TRAVEL
ROLL TRAVEL
PITCH TRAVEL
1. PITCH AND ROLL TRIM SWITCH
2. 3-AXIS-CONTROL STICK
3. SELF-CONTAINED ELECTRONICS
4. COMMUNICATIONS SWITCH
5. BACKUP FLIGHT CONTROL SYSTEM ENGAGE SWITCH

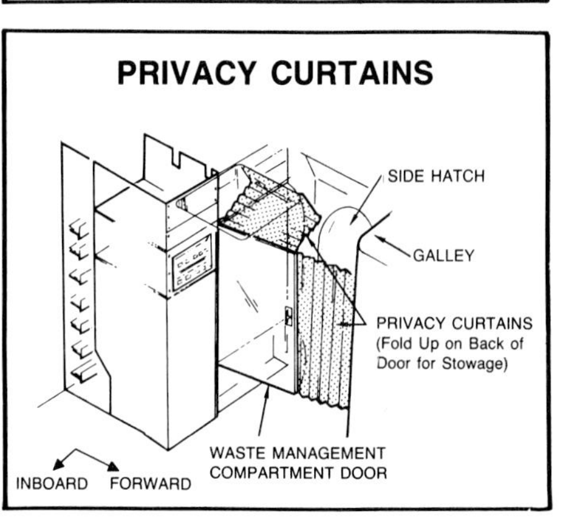
PRIVACY CURTAINS
SIDE HATCH
GALLEY
PRIVACY CURTAINS (Fold Up on Back of Door for Stowage)
WASTE MANAGEMENT COMPARTMENT DOOR
INBOARD
FORWARD

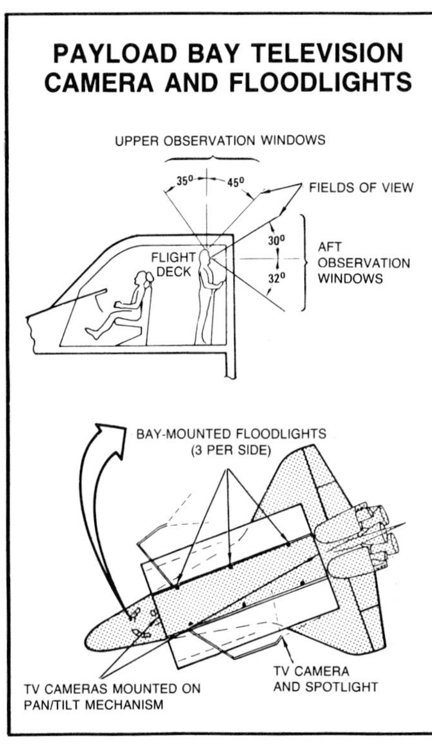
PAYLOAD BAY TELEVISION CAMERA AND FLOODLIGHTS
UPPER OBSERVATION WINDOWS
35°
45°
FIELDS OF VIEW
FLIGHT DECK
30°
32°
AFT OBSERVATION WINDOWS
BAY-MOUNTED FLOODLIGHTS (3 PER SIDE)
TV CAMERAS MOUNTED ON PAN/TILT MECHANISM
TV CAMERA AND SPOTLIGHT

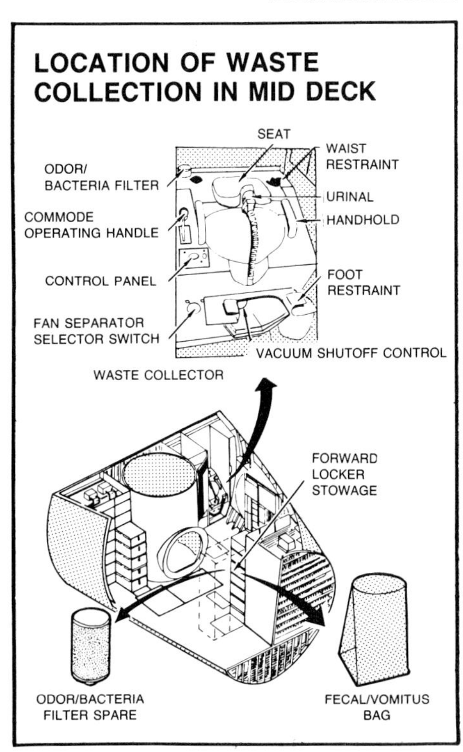
LOCATION OF WASTE COLLECTION IN MID DECK
SEAT
WAIST RESTRAINT
ODOR/ BACTERIA FILTER
URINAL
COMMODE OPERATING HANDLE
HANDHOLD
CONTROL PANEL
FOOT RESTRAINT
FAN SEPARATOR SELECTOR SWITCH
VACUUM SHUTOFF CONTROL
WASTE COLLECTOR
FORWARD LOCKER STOWAGE
ODOR/BACTERIA FILTER SPARE
FECAL/VOMITUS BAG

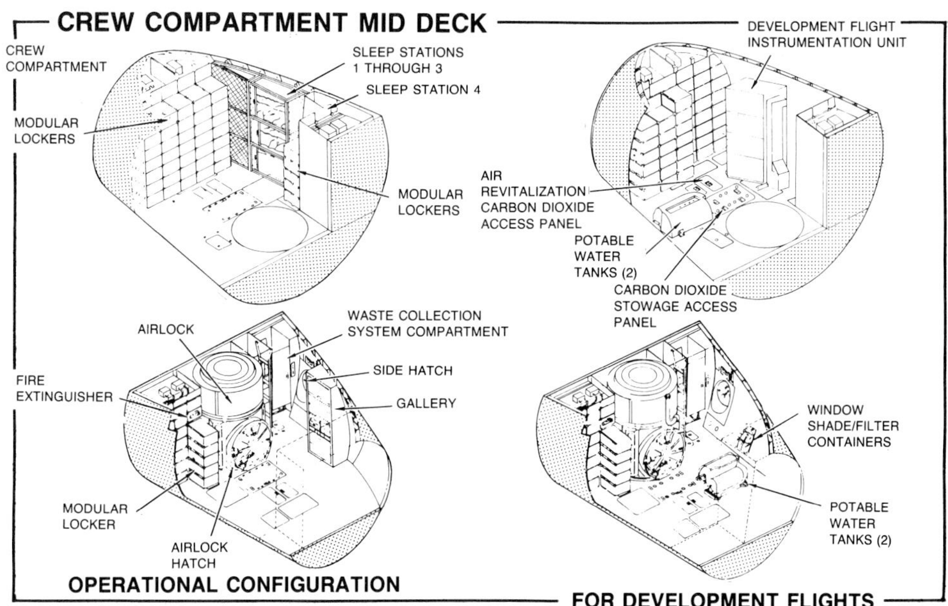

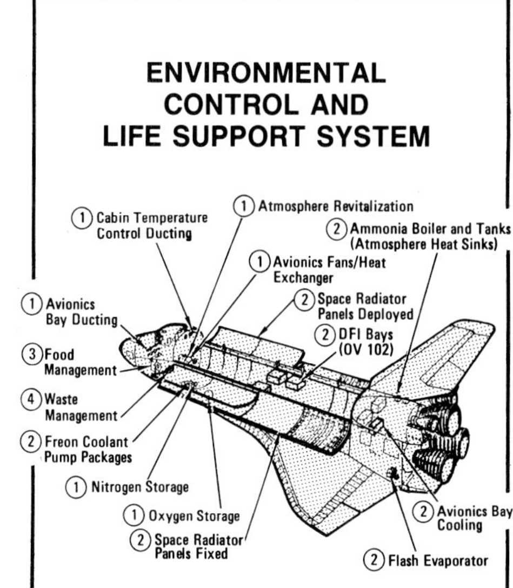

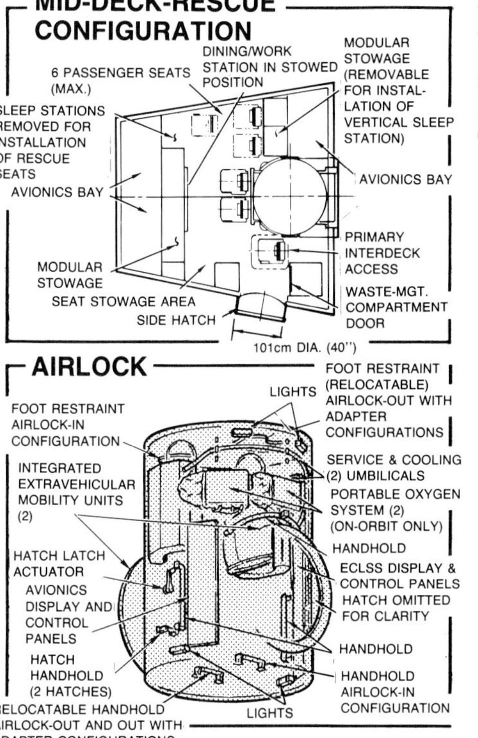

The airlock assembly for OV-105 is shown during 1987 at the Rockwell plant in Downey, California.

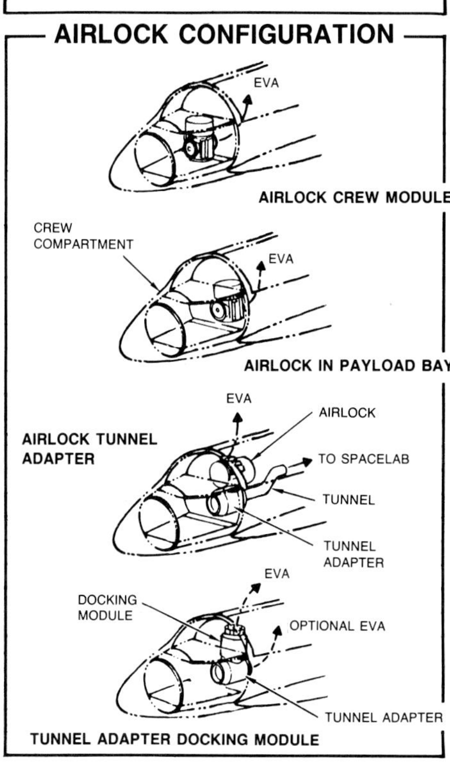

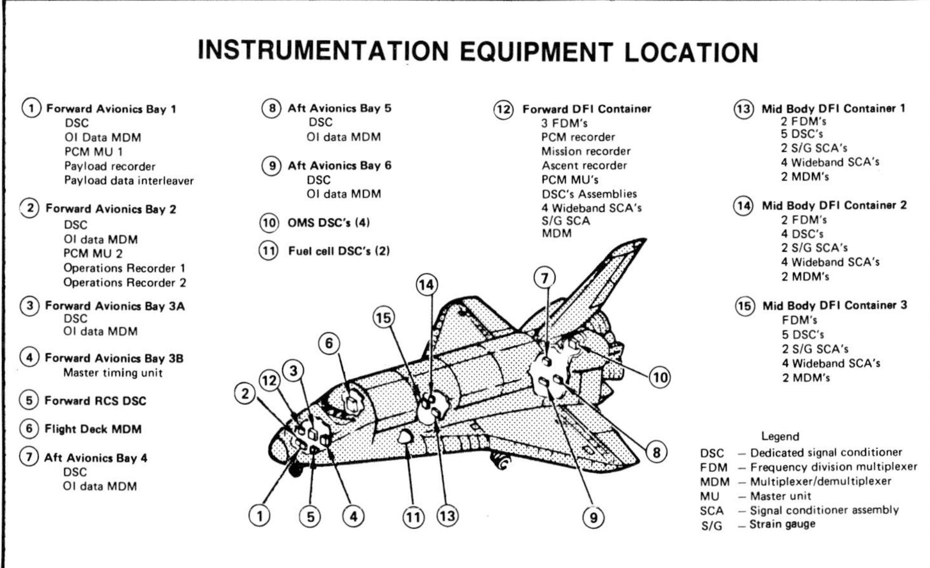

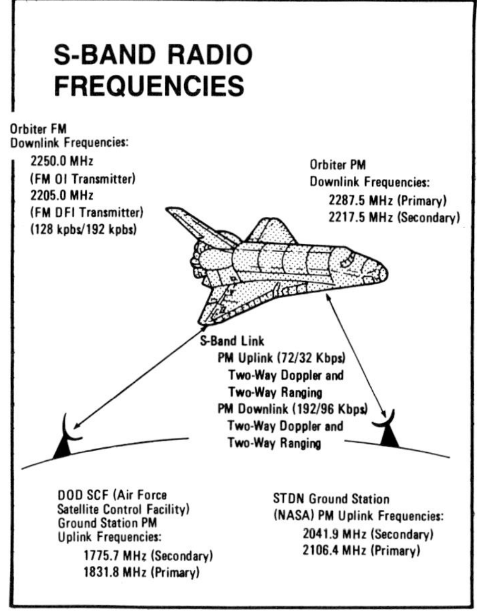

STS-1 crew, John Young and Robert Crippen, practice an emergency egress from a JSC training mock-up. The mock-up is transportable using the wheels seen at right.

"Columbia's" crew module before installation. Module was constructed at Rockwell's plant in Downey, California, then shipped to the assembly plant in Palmdale.

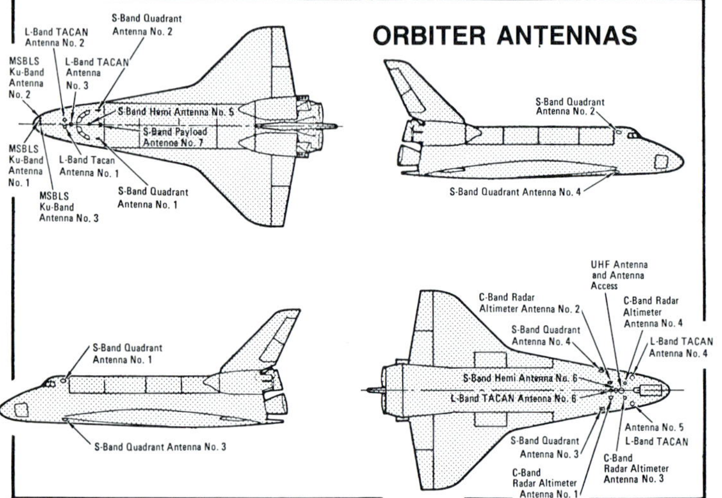

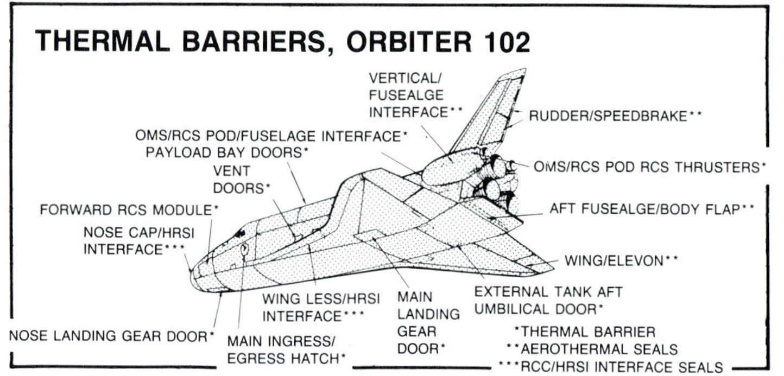

"Atlantis" crew module being lowered into forward fuselage at Palmdale. Flight deck windows have been installed, and the crew module wrapped with insulation.

Crew module for OV-105 being lowered into place during a fit-check to verify tolerances. Location of 40-inch crew egress hatch is visible to left.

Rockwell via Gerald Balzer

"Enterprise" crew module being mated. Due to lack of requirement, "Enterprise" was not thermal insulation equipped. Openings in upper crew module permitted ejection.

Rockwell via Erik Simonsen

"Atlantis" forward fuselage section at Palmdale prior to the installation of the forward RCS. The forward RCS contains 14 primary and 2 vernier thrusters.

NASA

"Discovery's" forward RCS is removed for servicing in the Orbiter Processing Facility at KSC. System provides attitude control during space flight.

RCC WING LEADING EDGE, ORBITER 102
(AND POSSIBLY SOME OF ORBITER 099)

LEADING EDGE RCC PANELS (22 L and 22 R)
A-286 BOLTS AND INCONEL BUSHINGS
HRSI
RCC SEAL STRIPS (22 L and 22 R)
COATED RCC LEADING EDGE PANEL
LEADING EDGE SPAR
Inconel 718 Fittings
INTERNAL INSULATION
COATED RCC TEE SEAL
RCC WING PANEL
RCC TEE SEAL
A-286 BOLTS AND INCONEL BUSHINGS
HRSI
FLOW RESTRICTOR SEAL

RCC WING LEADING EDGE
(POSSIBLY SOME OF ORBITER 099, ORBITER 103 & SUBSEQUENT)

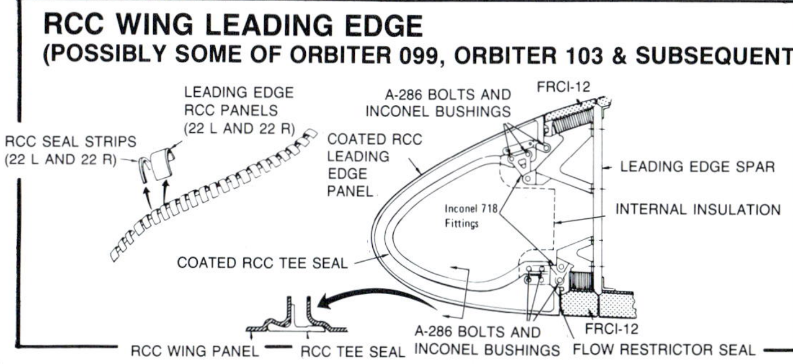

RCC NOSE CAP, ORBITER 102
(POSSIBLY SOME OF ORBITER 099)

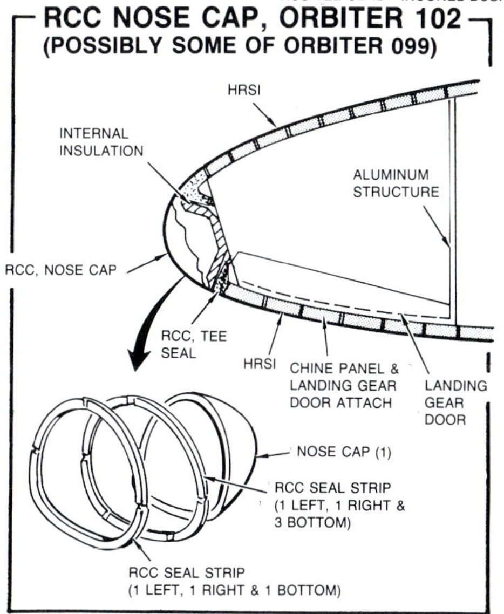

RCC NOSE CAP
(POSSIBLY SOME OF ORBITER 099, ORBITER 103 AND SUBSEQUENT)

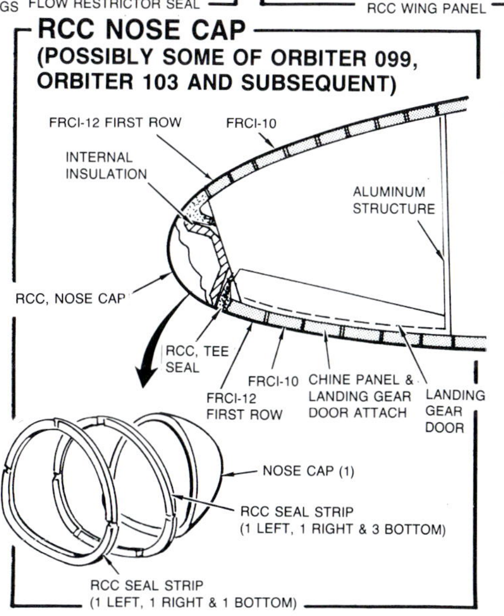

NASA via Erik Simonsen

The RCC nose cap of "Columbia" following STS-3. Small pitot tubes on each nose side are noteworthy.

NASA via Gerald Balzer

"Challenger" mid-payload bay looking forward. Airlock hatch is visible in open position.

Rockwell via Gerald Balzer

OV-101 nearing completion at Rockwell's Palmdale facility. Ejection seat panels are visible on cabin roof.

NASA

"Challenger" served as the structural test article. Later it would be utilized for static load and thermal tests.

NASA

OV-105 mid-fuselage (60 ft. long, 17 ft. wide, 13 ft. high and weighing 13,500 lbs.), built by General Dynamics in San Diego, sits on jack stands at Palmdale.

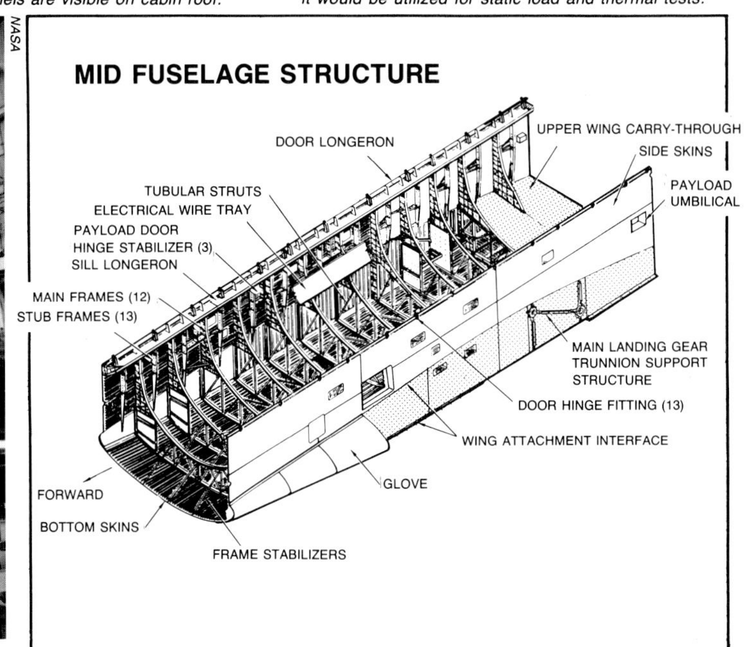

Rockwell via Erik Simonsen

"Discovery" during assembly at Palmdale. At this point the orbiter forward and mid-fuselages have been mated with the wings.

Rockwell via Erik Simonsen

"Atlantis" with all major components, including the crew cabin and vertical tail, mated. Forward sections of payload bay doors still are missing to provide payload bay access.

"Challenger" nearing completion after modifications from structural test article (STA-099) to OV-099. Flat surface is where right OMS pod will be mounted.

Aft end of "Columbia" shows cut-outs for three SSMEs. The body-flap and elevons still need to be fitted. Little thermal tile work has been done.

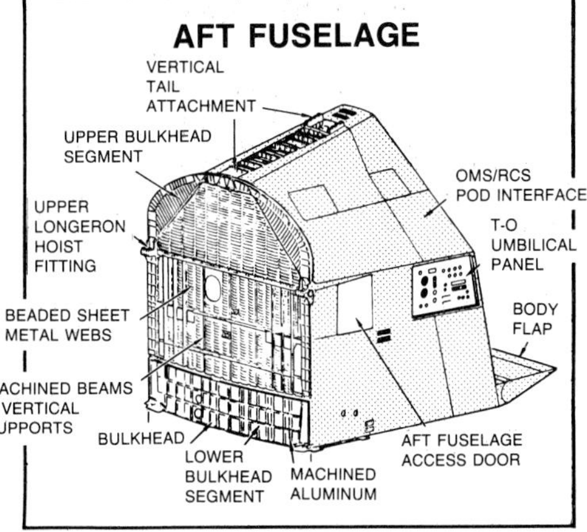

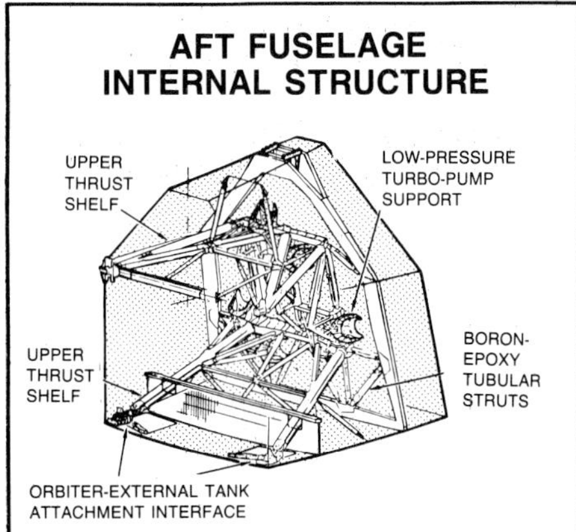

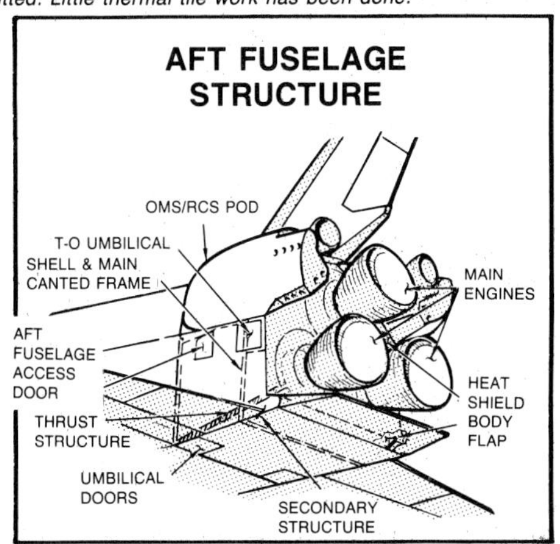

OSTA-1 pallet, for earth survey work, installed in "Columbia" for STS-2. Visible are shiny radiator surfaces and the stowed RMS along the left longeron. Foil-covered box in the middle of RMS is a TV camera.

Looking down and forward on the payload bay of "Enterprise". There were no radiators fitted to OV-101.

"Columbia's" payload bay before completion. Payload bay doors are shut, but no radiators are visible.

Rockwell via Erik Simonsen

Remote manipulator arm in the payload bay of "Columbia". RMS (50 ft. long) is built by Spar Aerospace in Canada. Black squares are pressure equalization vents.

Rockwell via Erik Simonsen

OV-103 during roll-out at Palmdale on October 16, 1983. Extensive use of flexible Nomex blankets is evident. Blankets are easier to install and maintain than RSI tiles.

NASA via Gerald Balzer

View of "Columbia" payload bay during STS-1. A few white tiles are missing from left OMS pod.

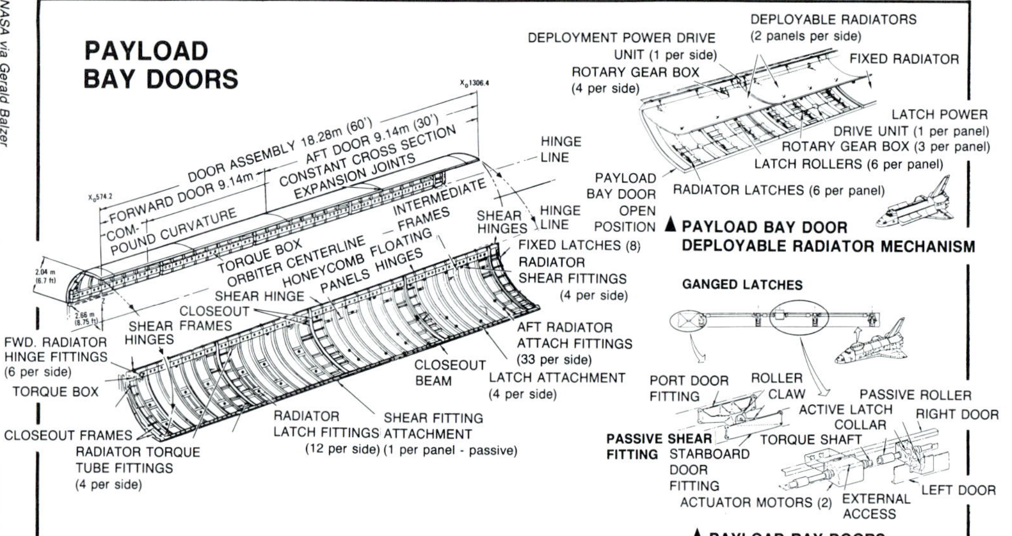

Rockwell

"Challenger's" payload bay doors during final assembly at Palmdale. Each graphite-epoxy/Nomex composite door is 60 ft. long and weighs 1,632 lbs.

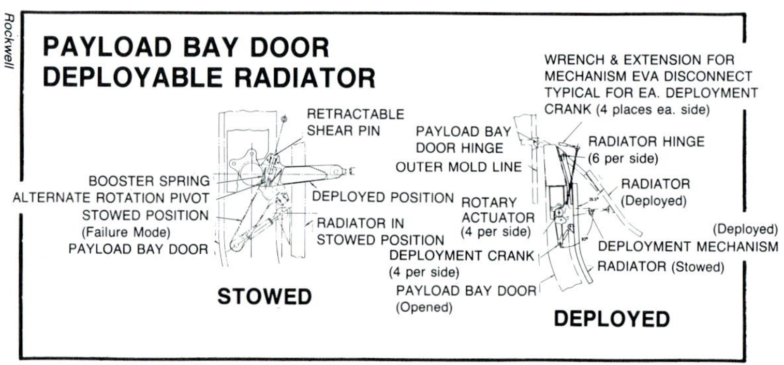

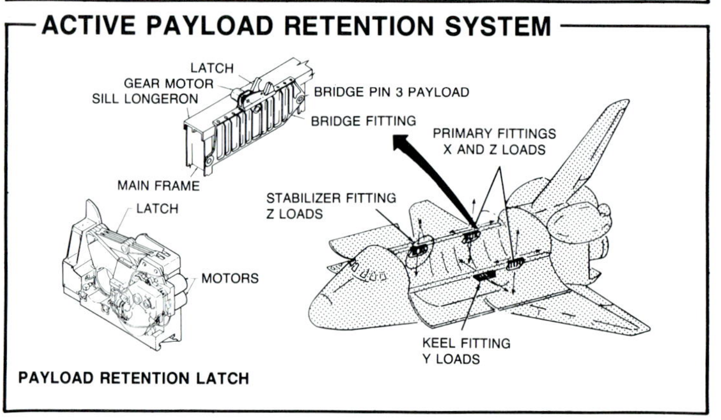

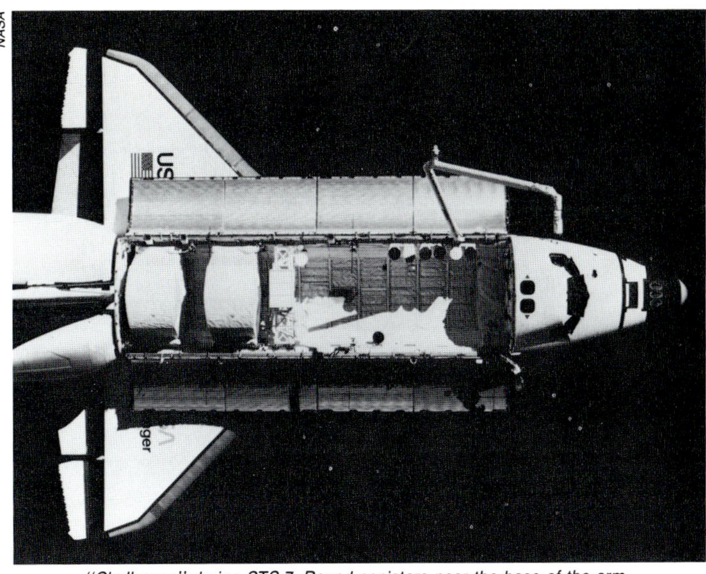

"Discovery's" remote manipulator arm during STS-19 (51-A) in which two stranded satellites were recovered. Various assemblies can be fitted to the end of the arm.

"Challenger" during STS-7. Round canisters near the base of the arm, house "get-away-special" (GAS) payloads.

MECHANICAL ARM
STOWED POSITION AND GENERAL ARRANGEMENT

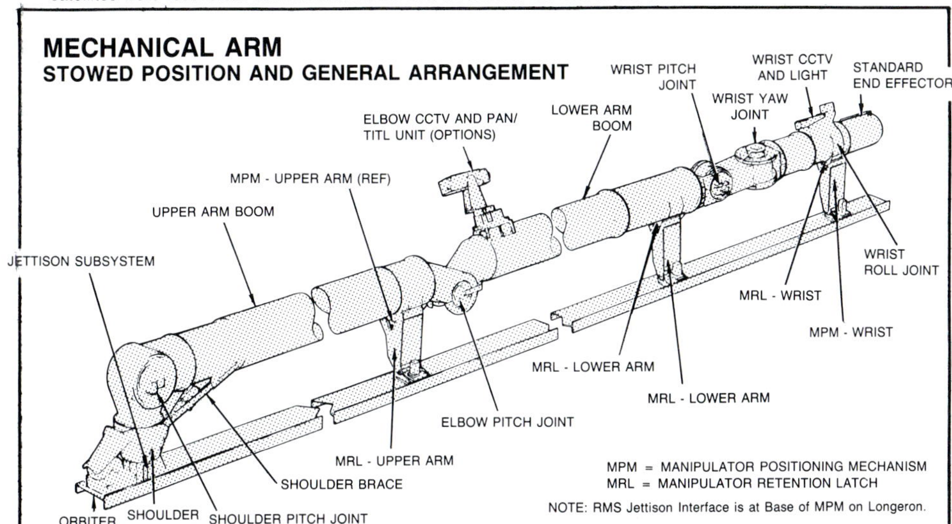

RMS ROTATION HAND CONTROL SWITCHES

1. RATE HOLD (2-POSITION PUSHBUTTON)
2. RATE LIMIT VERNIER/COARSE (2-POSITION SLIDE SWITCH)
3. END EFFECTOR CAPTURE/RELEASE (3-POSITION ROCKER)

SNARE CAPTURE AND REGIDIZATION SEQUENCE

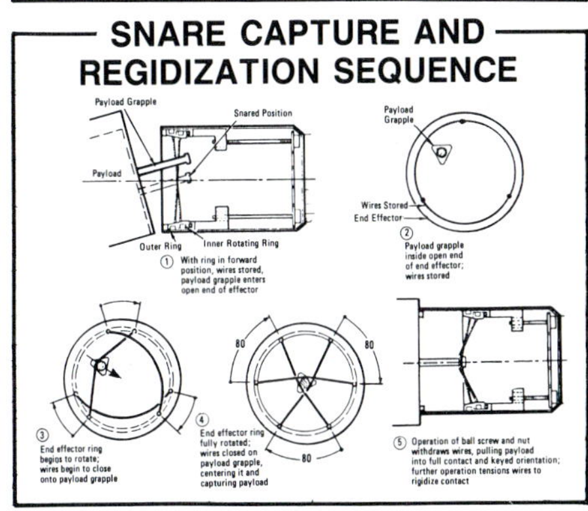

RMS COMPONENTS

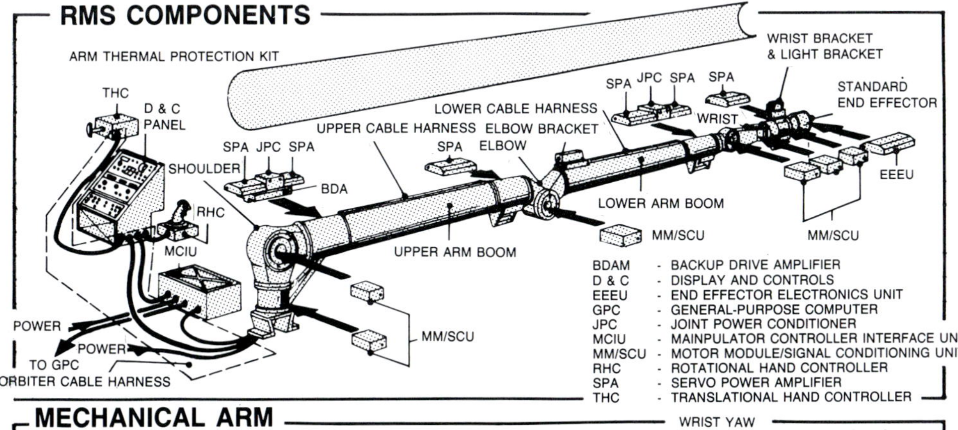

BDAM	-	BACKUP DRIVE AMPLIFIER
D & C	-	DISPLAY AND CONTROLS
EEEU	-	END EFFECTOR ELECTRONICS UNIT
GPC	-	GENERAL-PURPOSE COMPUTER
JPC	-	JOINT POWER CONDITIONER
MCIU	-	MAINPULATOR CONTROLLER INTERFACE UNIT
MM/SCU	-	MOTOR MODULE/SIGNAL CONDITIONING UNIT
RHC	-	ROTATIONAL HAND CONTROLLER
SPA	-	SERVO POWER AMPLIFIER
THC	-	TRANSLATIONAL HAND CONTROLLER

MECHANICAL ARM
STOWED POSITION AND MOVEMENT CONFIGURATION

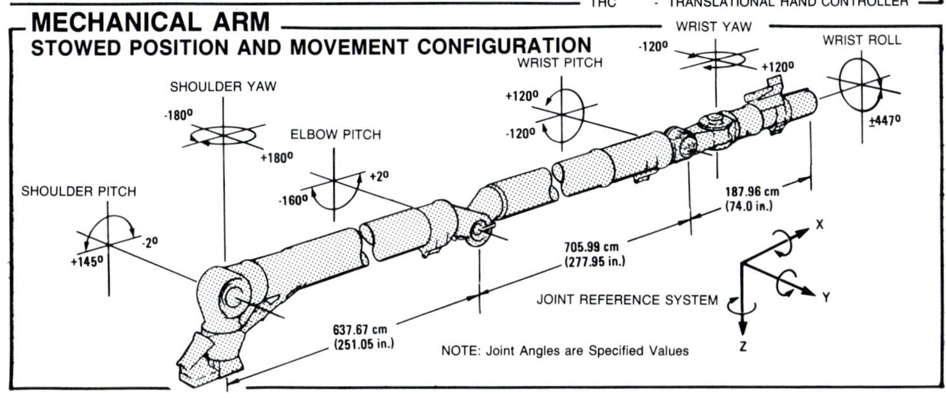

OVERALL CONFIGURATIONS

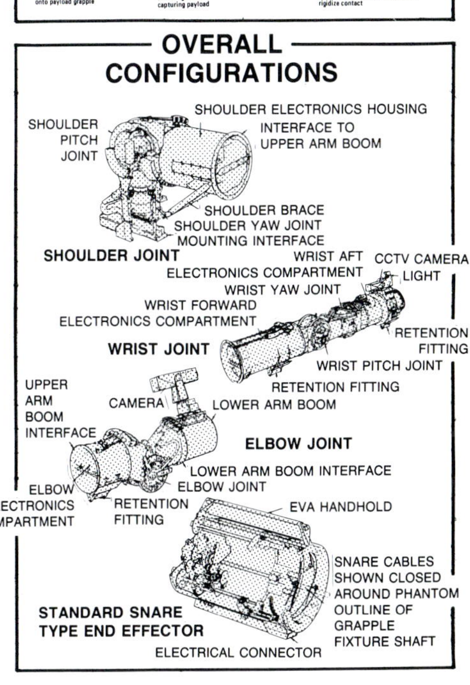

NASA

OV-105 right wing about to be mated to mid-fuselage section. OV-105 major pieces originally were built as structural "spares" set during 1983/84.

NASA

Shuttle wings, built by Grumman Aerospace in New York, were transported by ship to California. Each was 60 ft. long and 5 ft. thick (at the fuselage intersection).

WING CONFIGURATION

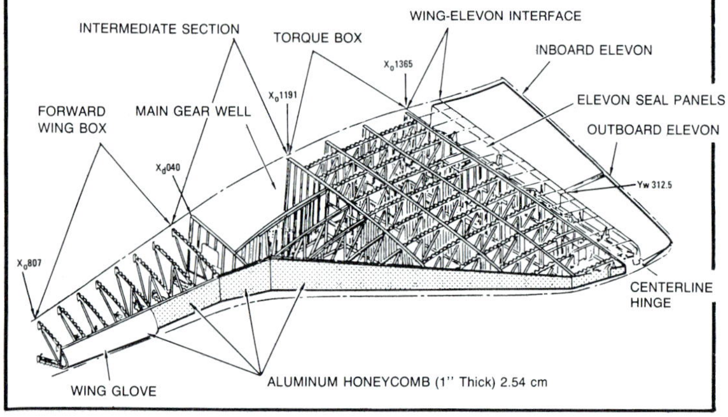

INTERMEDIATE SECTION

TORQUE BOX

WING-ELEVON INTERFACE

INBOARD ELEVON

X_o1365

X_o1191

ELEVON SEAL PANELS

OUTBOARD ELEVON

FORWARD WING BOX

MAIN GEAR WELL

X_o1040

Yw 312.5

X_o807

CENTERLINE HINGE

WING GLOVE

ALUMINUM HONEYCOMB (1" Thick) 2.54 cm

AERODYNAMIC SURFACES

RUDDER/SPEED BRAKE — 2 VERTICAL PANELS
- Yaw control when both panels deflected left or right (rudder function)
- Aerodynamic drag control when panels opened (away from each other) or closed (toward each other) (speed brake function)

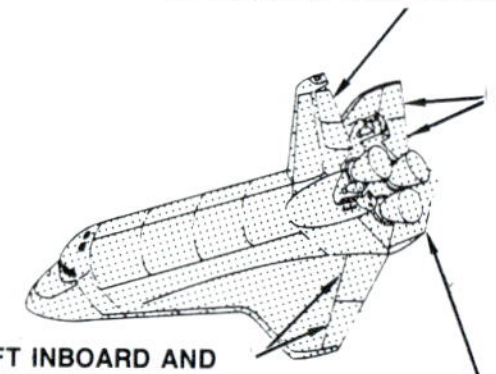

RIGHT INBOARD AND OUTBOARD ELEVONS
- Pitch control when deflected same direction as left elevons (elevator function)
- Roll control when deflected opposite direction of left elevons (aileron function)
- Inboard and outboard deflected identically during entry.

LEFT INBOARD AND OUTBOARD ELEVONS

BODY FLAP
- Main engine thermal protection
- Pitch trim to reduce elevon deflections

Rockwell via Gerald Balzer

The two-section elevons of "Enterprise" during assembly. Foam substitutes for the tiles visible in this view are quite thick.

NASA/LTV

Underside of an orbiter wing shows the intricate tile work. Light gray leading edge is constructed of Vought Corp. manufactured carbon-carbon.

Rockwell via Erik Simonsen

"Columbia" in the mate-demate device at Dryden FRC after STS-4. The tile work around the outboard elevon and wingtip are readily visible.

Rockwell via Erik Simonsen

Columbia shows TPS tile detail around the main landing gear door after STS-4. The carbon-carbon leading edge can withstand heat up to 3,000° F.

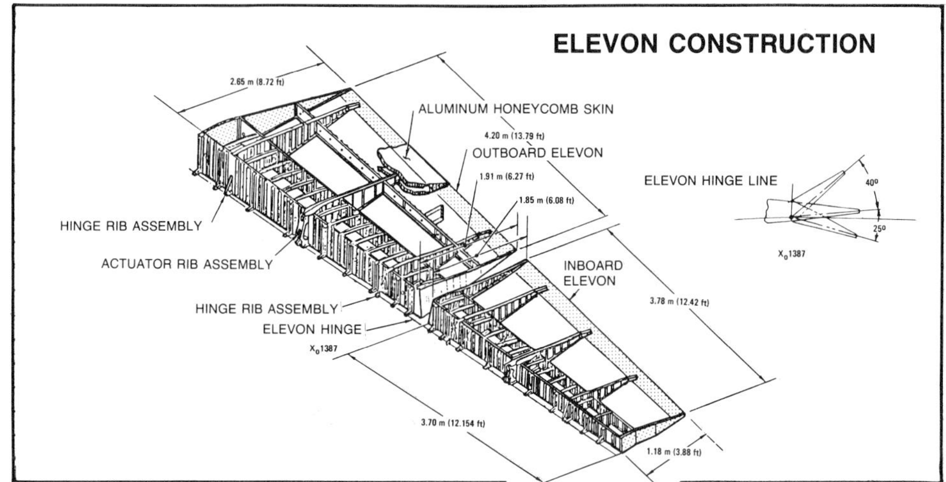

Rockwell via Erik Simonsen

View showing the demarcation line between white tiles and blankets just aft of the wing leading edge.

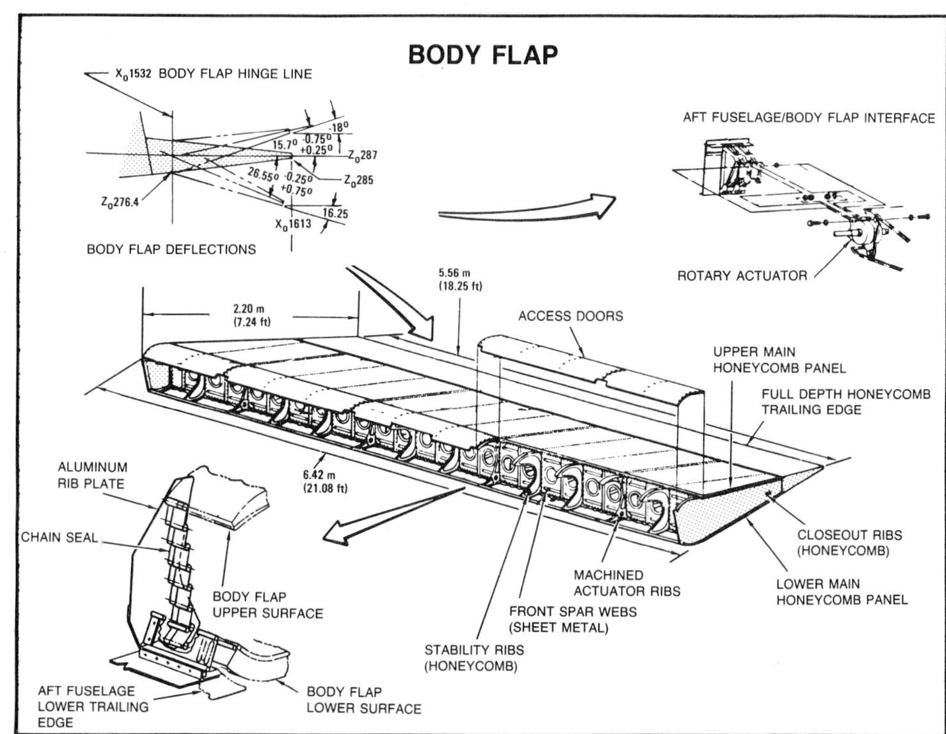

Rockwell via Erik Simonsen

Base of vertical fin has TPS in place to protect from engine exhaust heat rather than reentry heat.

Rockwell via Erik Simonsen

Mating of "Discovery" vertical stabilizer to aft fuselage at Palmdale. Black area just forward of rudder is an Inconel honeycomb hinge seal.

NASA

Shuttle Infrared Leeside Temperature Sensing (SILTS) experiment pod (20 in. dia.) was added on top of "Columbia's" vertical stabilizer during 1984/85 overhaul.

Lower aft section of "Enterprise" vertical stabilizer. Inconel hinge seal between the rudder/speedbrake and the vertical stabilizer is discernible.

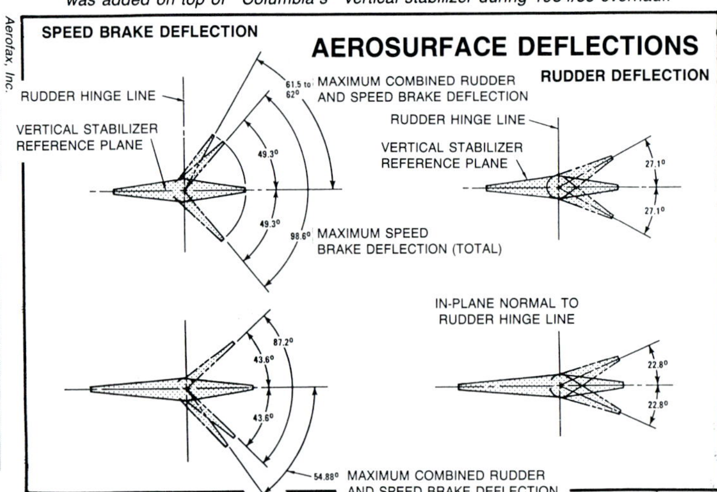

Aerofax, Inc.

NASA via Gerald Balzer

Right vertical fin assembly with split rudder and thermal tiles in place. Split rudder serves as deceleration device during approach to landing.

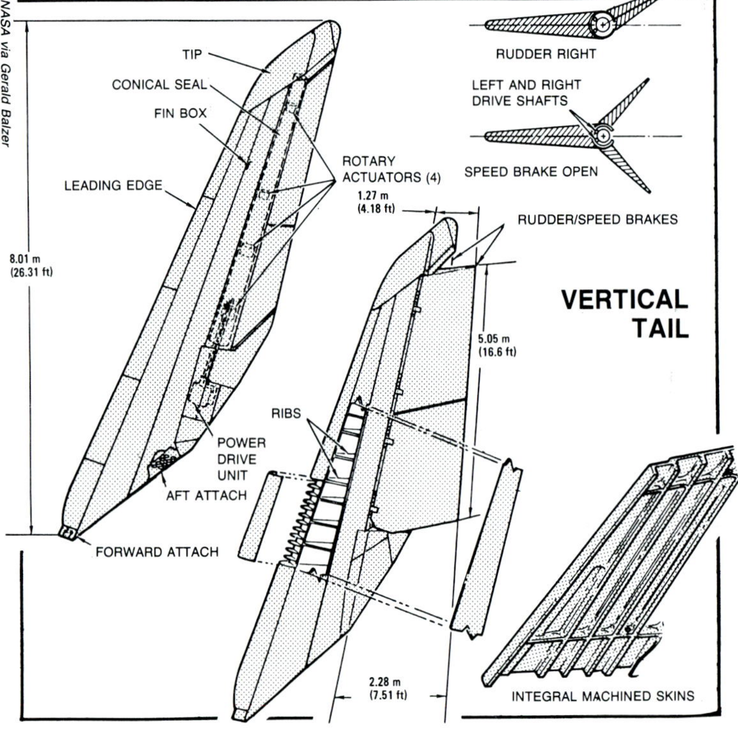

Aerofax, Inc.

"Enterprise" nose landing gear. OV-101 was not fitted with the nose-wheel steering.

Aerofax, Inc.

Rear view of Menasco-manufactured nose landing gear. Gear can not be retracted after extension.

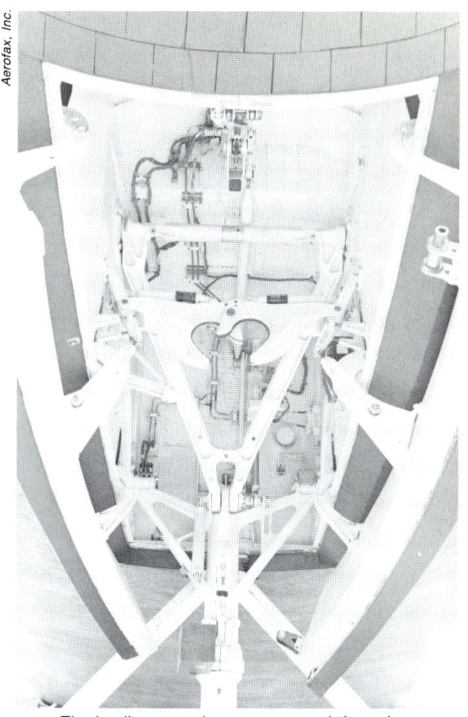

Aerofax, Inc.

The landing gear doors are opened through a mechanical linkage attached to large hooks.

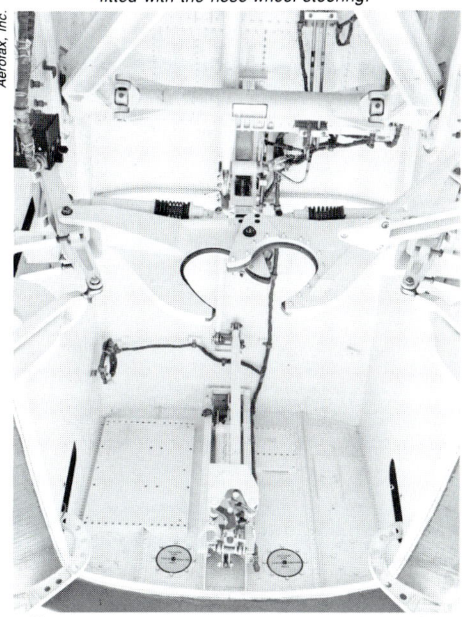

Aerofax, Inc.

Clamps grasp the gear strut when it is pushed up into the gear well. It is held in position by an up-lock hook.

Rockwell via Erik Simonsen

Only right door has up-locks. Hydraulic cylinder retracts gear, but can be actuated only on the ground.

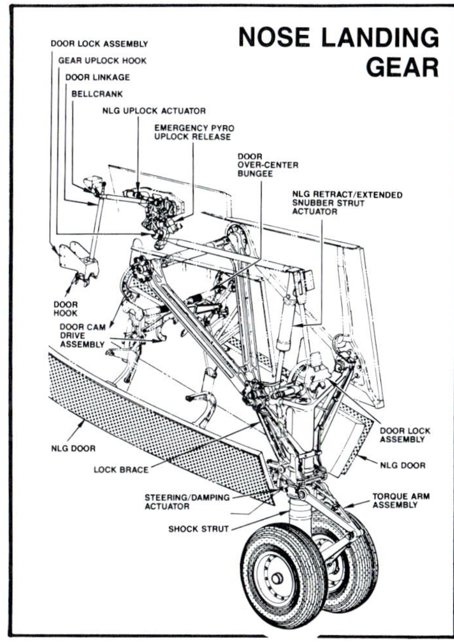

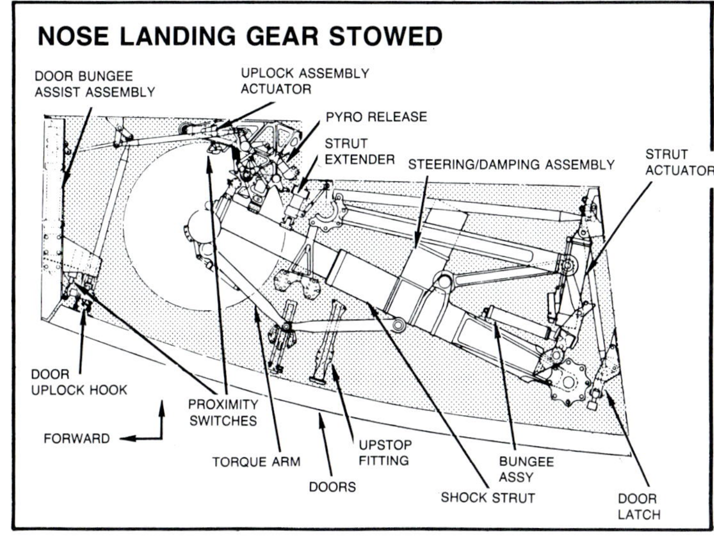

NASA via Gerald Balzer

The STS-23 (51-D) post-mission landing resulted in one blown and one severely damaged tire. This led to a decision to end all succeeding missions at Edwards AFB.

Sans brake hydraulic lines "Enterprise's" right main gear. Vehicle is in storage at Dulles Airport.

Like the nose gear, the main gear free-falls into position with some help from springs and hydraulics.

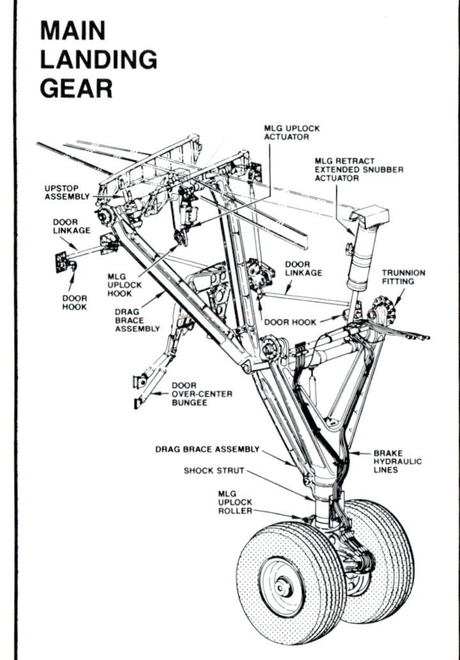

Aerofax, Inc.

Main gear doors are mechanically actuated by center brace. Brace rests on top of gear strut.

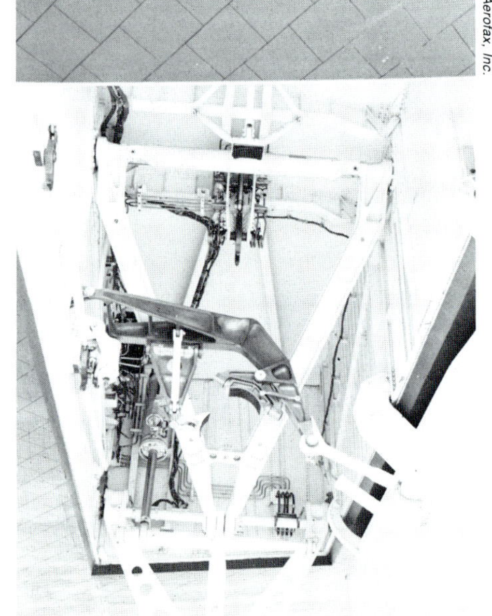

Left main gear mechanisms are mirror image of right. Hydraulic retraction actuator acts as snubber.

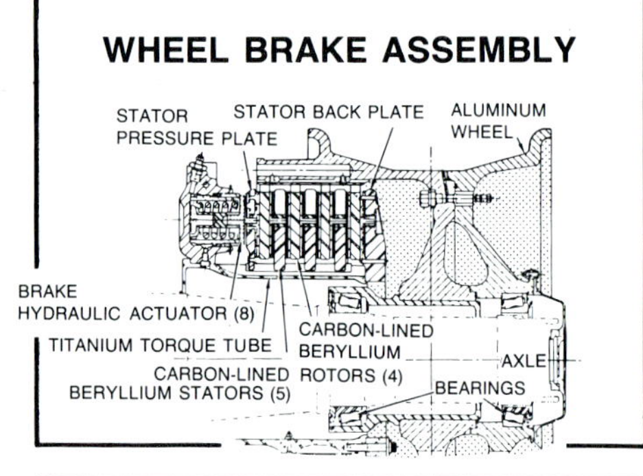

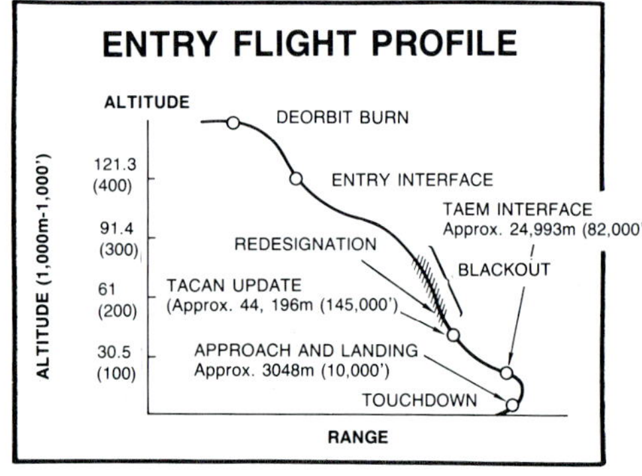

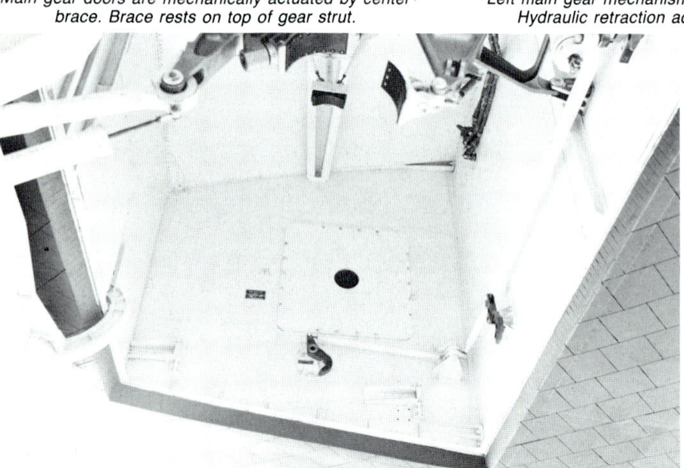

Forward wall of right main gear well is basically empty. Hooks along lower portion of well are up-locks for gear door.

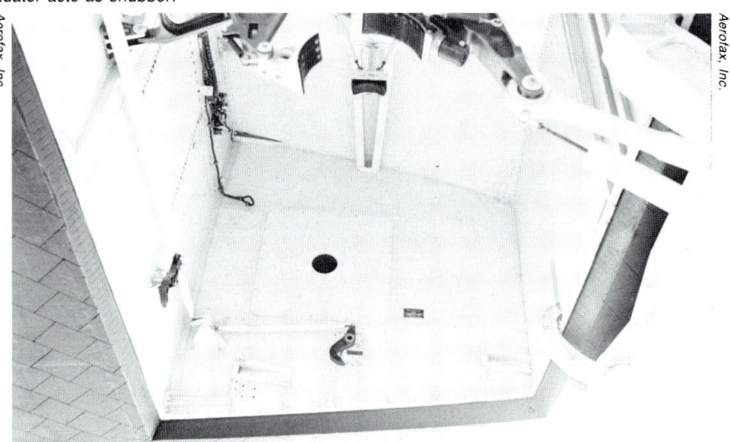

Left main gear well also is empty. A snubber is located in the upper center of the picture and serves as an up-stop for the gear strut.

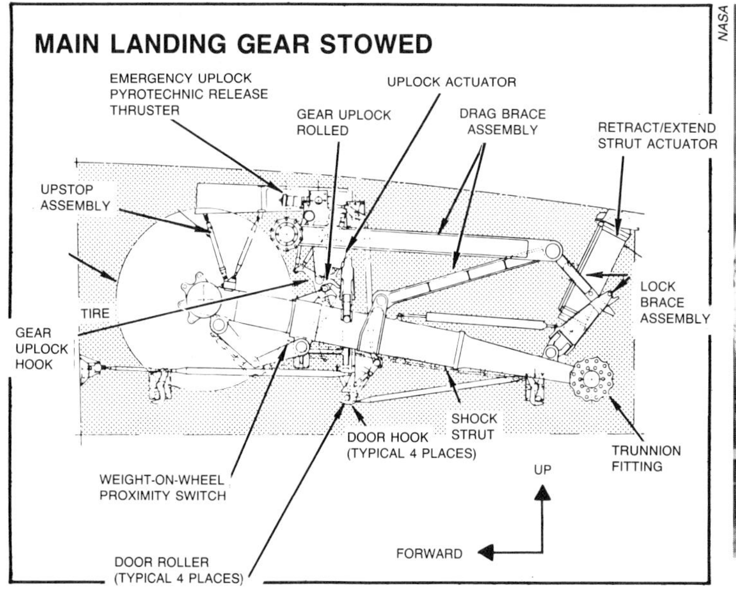

NASA

"Enterprise" during assembly illustrates some of the hydraulic brake lines normally appearing on the main gear struts. Gear door snubber is visible.

Rockwell via Erik Simonsen

High temperature black tiles cover gear well doors to protect gear and wing interior. Technician is seen heat-curing temporary tile fix on "Columbia" following STS-4.

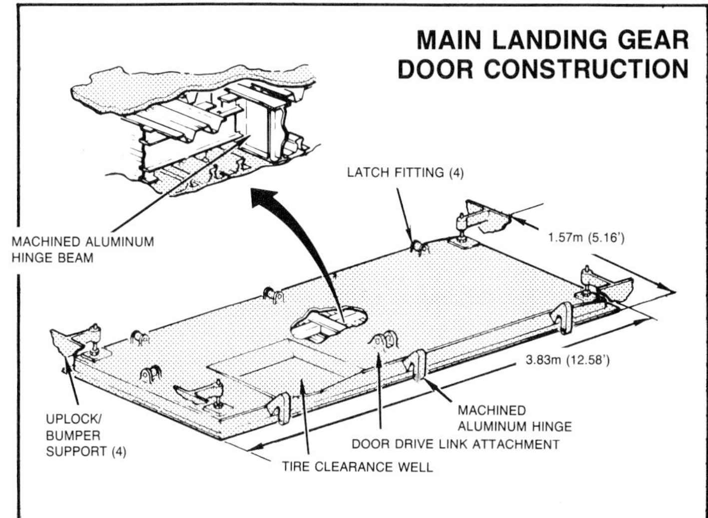

NASA via Erik Simonsen

Each SSME produces 375,000 lbs. of thrust at sea level. The engines are regeneratively cooled by liquid hydrogen flowing through thrust chamber walls.

NASA via Erik Simonsen

Covers protect the three SSMEs, two OMS engines, and the aft RCS. The latter coupled with the forward RCS provide vehicle attitude control.

Rockwell via Erik Simonsen

Each SSME consumes 889 lbs. of LO_2 and 146 lbs. of LH_2 per second at full thrust. High speed turbo-pumps feed propellants to the engines at high pressure. The SSME is the most sophisticated rocket engine presently available.

NASA via Erik Simonsen

Hydraulically gimballed SSMEs have maximum movement limits of ±10.5° in pitch and ±8.5° in yaw.

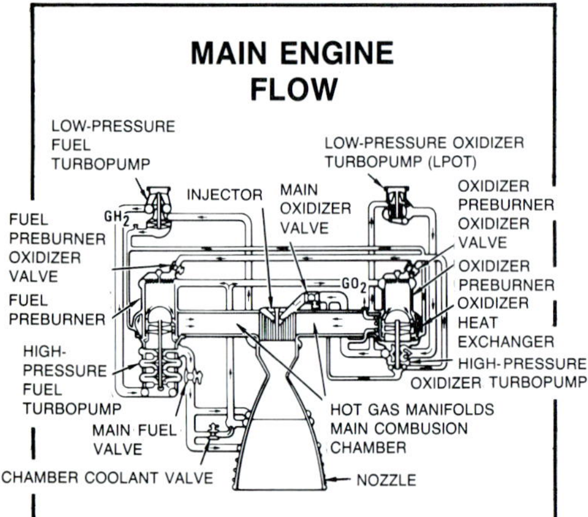

NASA via Gerald Balzer

First unpainted ET is readied for STS-3. This, and all subsequent, tanks were left unpainted to save 595 lbs. and $15,000. ET is manufactured by Martin Marietta in the same Michoud, Louisiana facility that produced Saturn V parts.

Martin Marietta via Erik Simonsen

Wedge nose of LO_2 tank reduces drag and serves as a lightning rod for the shuttle assembly during ascent.

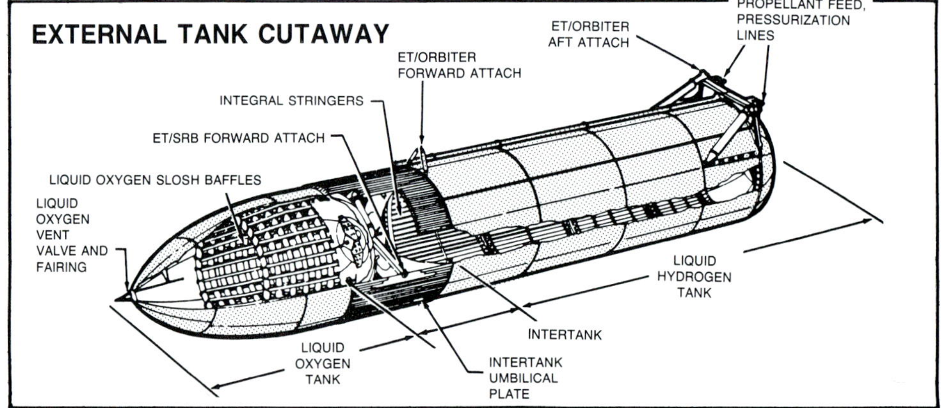

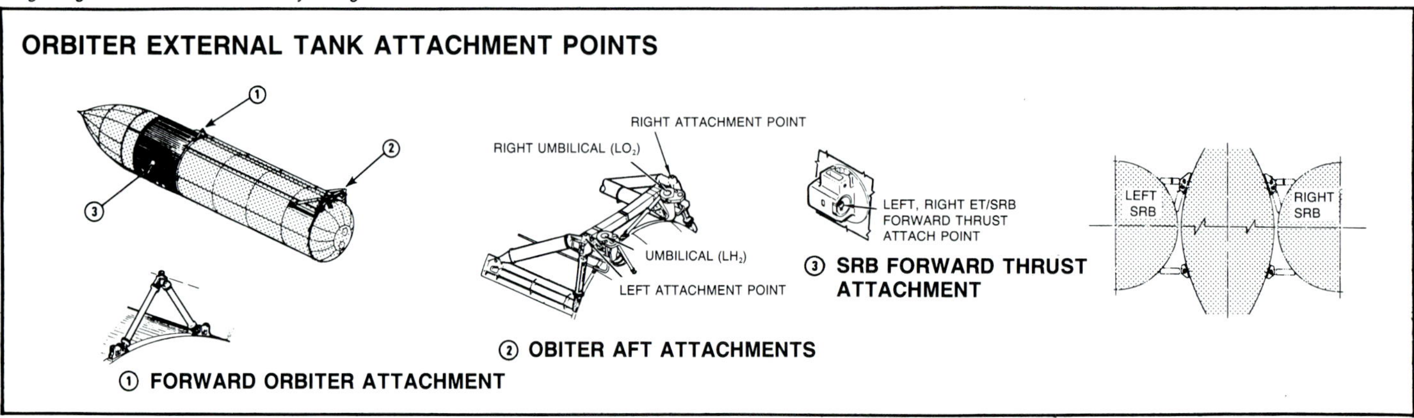

Martin Marietta via Erik Simonsen

Corrugated surface of the intertank separating the LO_2 and LH_2 tanks is apparent here. The external plumbing carries LO_2 to the 17 in. disconnects on the orbiter.

Dennis R. Jenkins

Aft attach struts secure the orbiter to the ET. "Pathfinder" external tank was part of the main propulsion test article (MPTA-ET) until it was retired for display.

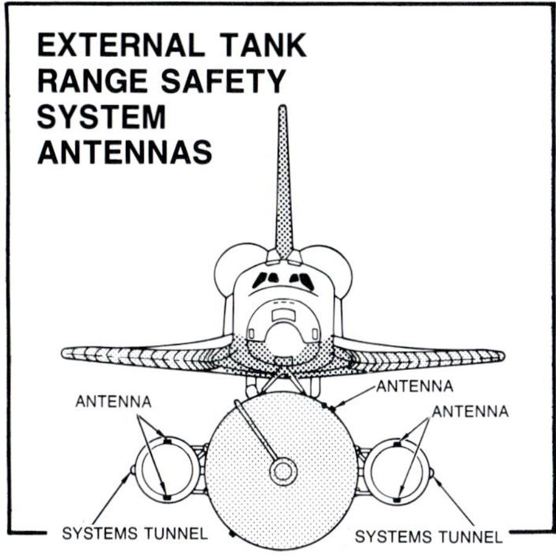

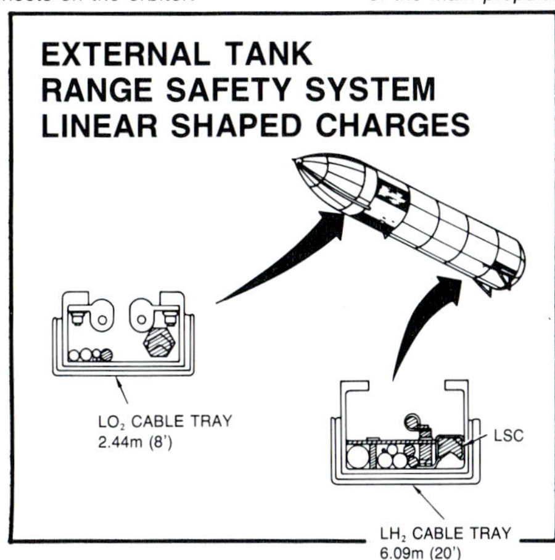

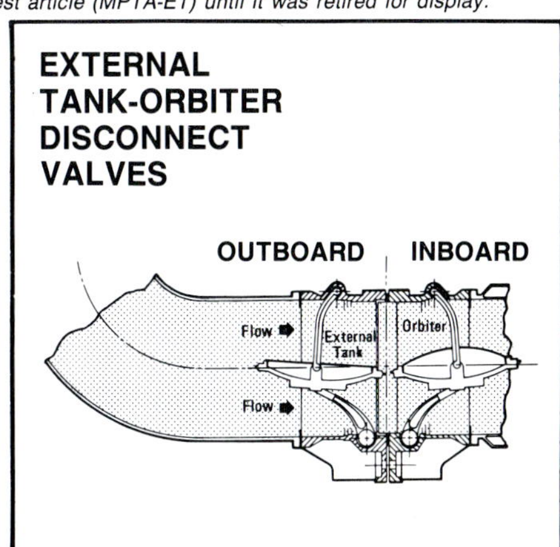

NASA

One of the OMS pods that would be used on "Discovery's" return-to-flight mission (STS-26R) is transported to a checkout cell at the Hypergolic Maintenance Facility. This pod makes extensive use of thermal blankets.

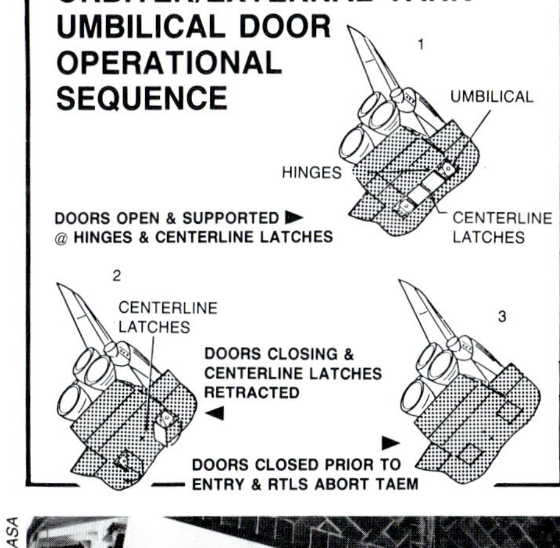

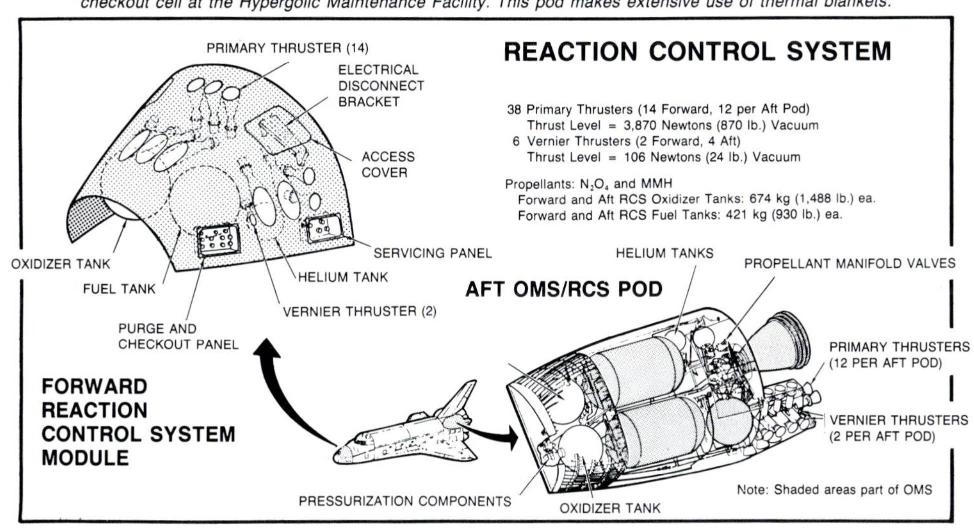

NASA

"Columbia's" OMS pod just before STS-2 showing nine of the twelve primary RCS thrusters.

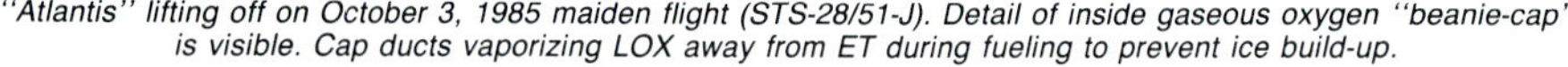

NASA

"Atlantis" lifting off on October 3, 1985 maiden flight (STS-28/51-J). Detail of inside gaseous oxygen "beanie-cap" is visible. Cap ducts vaporizing LOX away from ET during fueling to prevent ice build-up.

NASA via Erik Simonsen

Nose cone contains electronics, recovery aids and parachutes, as well as 4 solid fuel separation rockets.

NASA

For fit-check, OV-101 is rolled out of the VAB on May 1, 1979 with a pair on inert SRBs and an ET.

NASA via Gerald Balzer

Four separation motors are attached to each SRB aft skirt. Each motor is 31.1 in. long and 12.8 in. in dia. and provides thrust for 1.02 seconds to ensure ET/SRB separation.

Dennis R. Jenkins

Wedge-shaped aft skirt areas where 3.5 in. dia. "hold-down" bolts attach SRBs to MLP. Special filament wound casings (seen) were developed for Vandenberg launches.

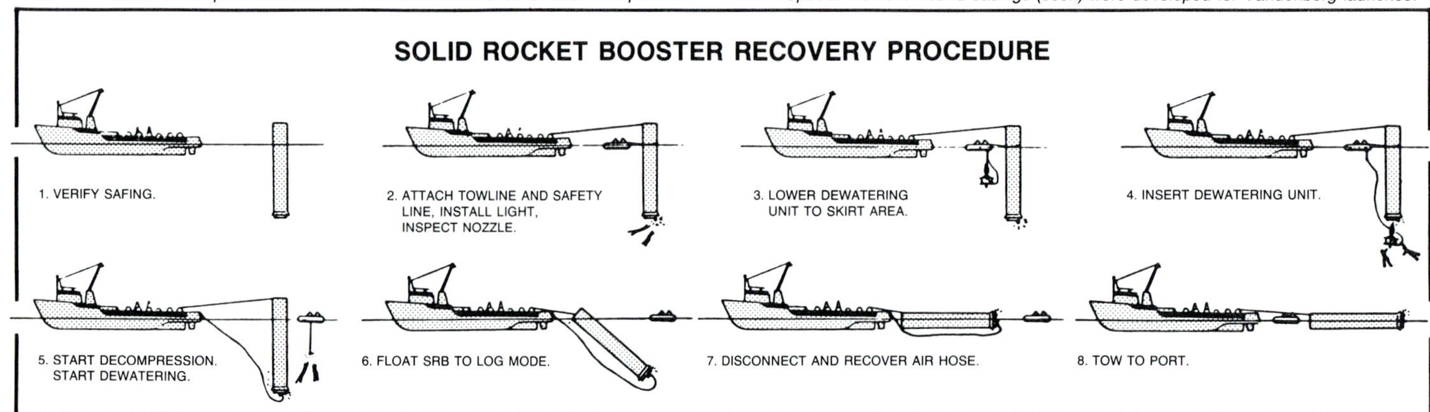

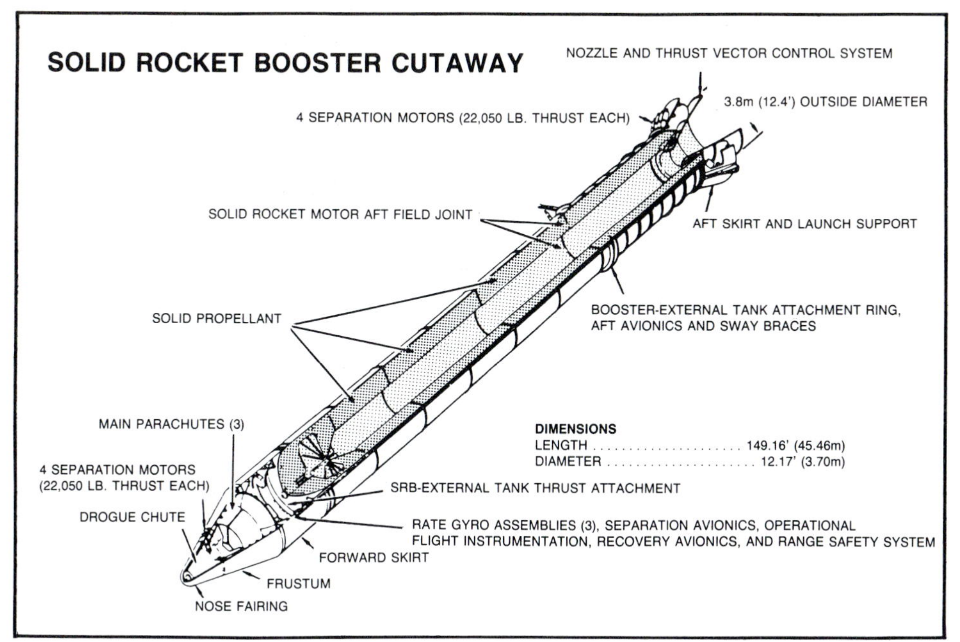

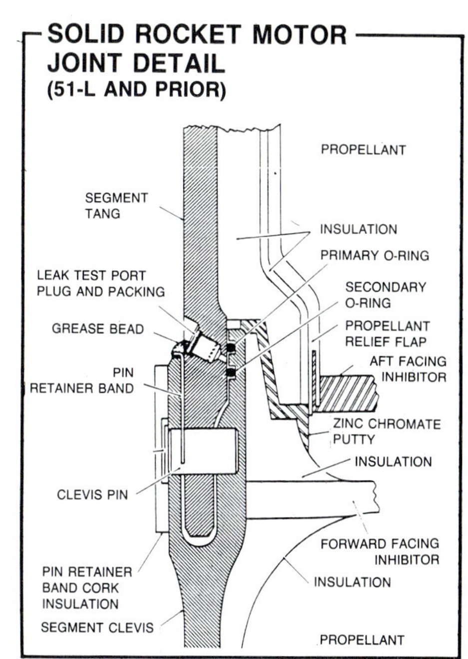

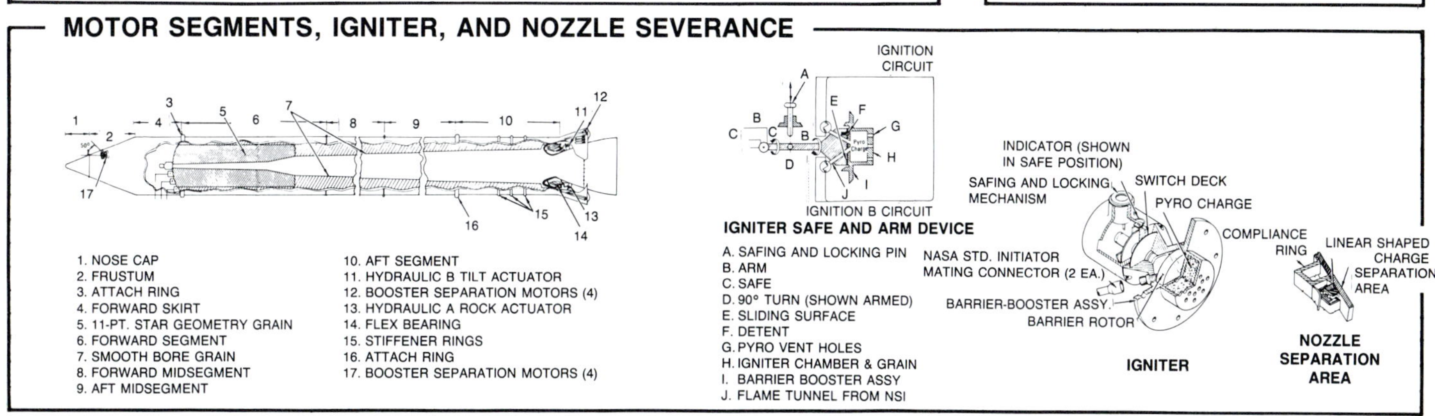

SEPARATION SYSTEM ELEMENTS

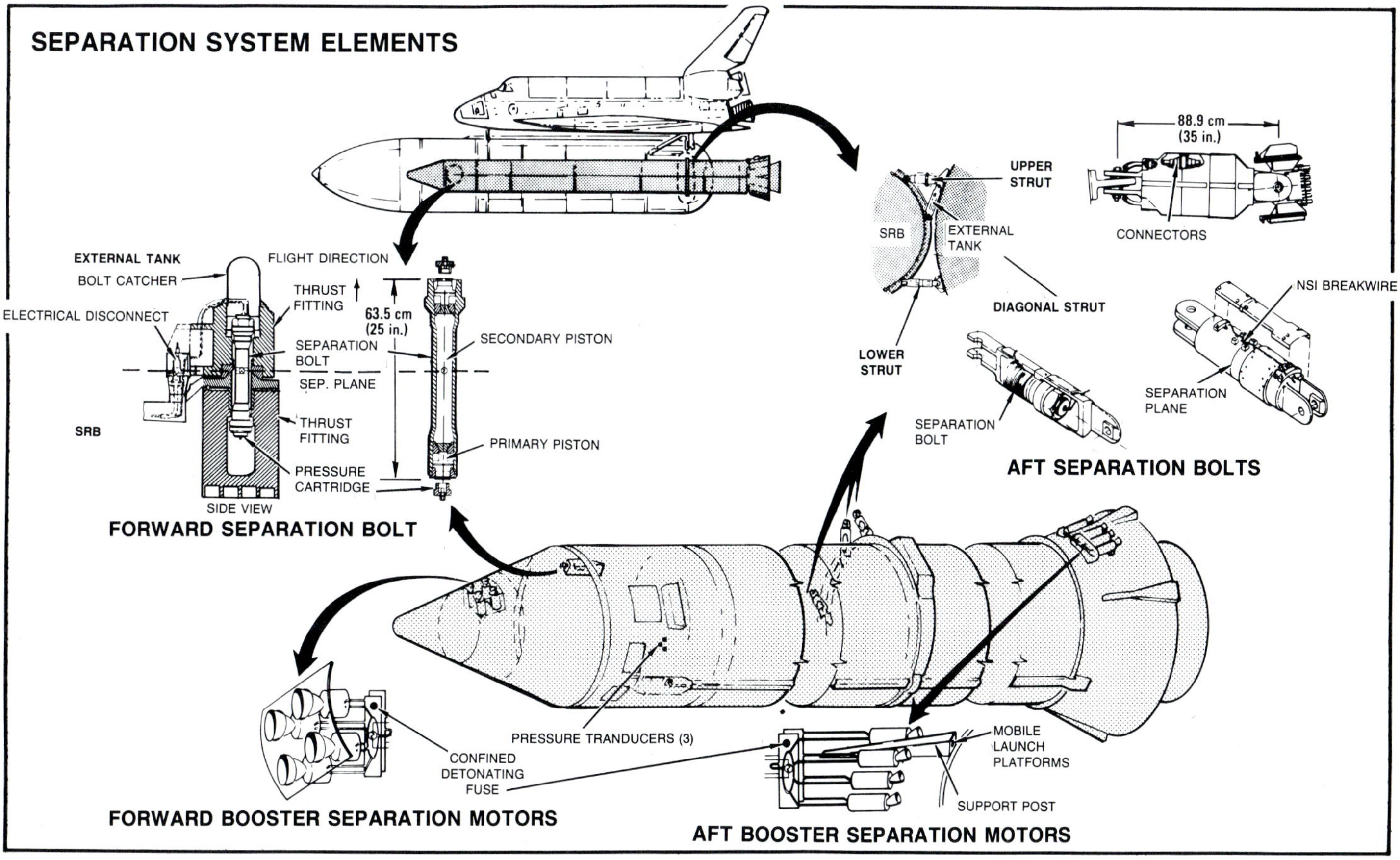

EXTERNAL TANK
BOLT CATCHER
FLIGHT DIRECTION
THRUST FITTING
ELECTRICAL DISCONNECT
63.5 cm (25 in.)
SEPARATION BOLT
SECONDARY PISTON
SEP. PLANE
SRB
THRUST FITTING
PRIMARY PISTON
PRESSURE CARTRIDGE
SIDE VIEW

FORWARD SEPARATION BOLT

UPPER STRUT
SRB
EXTERNAL TANK
DIAGONAL STRUT
LOWER STRUT
88.9 cm (35 in.)
CONNECTORS
NSI BREAKWIRE
SEPARATION PLANE
SEPARATION BOLT

AFT SEPARATION BOLTS

PRESSURE TRANDUCERS (3)
CONFINED DETONATING FUSE
MOBILE LAUNCH PLATFORMS
SUPPORT POST

FORWARD BOOSTER SEPARATION MOTORS

AFT BOOSTER SEPARATION MOTORS

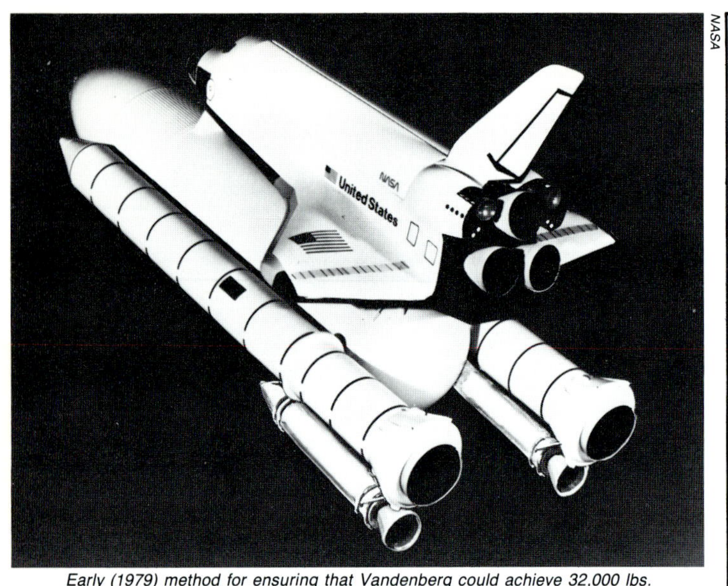

Early (1979) method for ensuring that Vandenberg could achieve 32,000 lbs. into polar orbit. Strap-on auxiliary SRMs would not be recovered.

ORBITER AIRBREATHING ENGINES AIRBREATHING PROPULSION SYSTEM

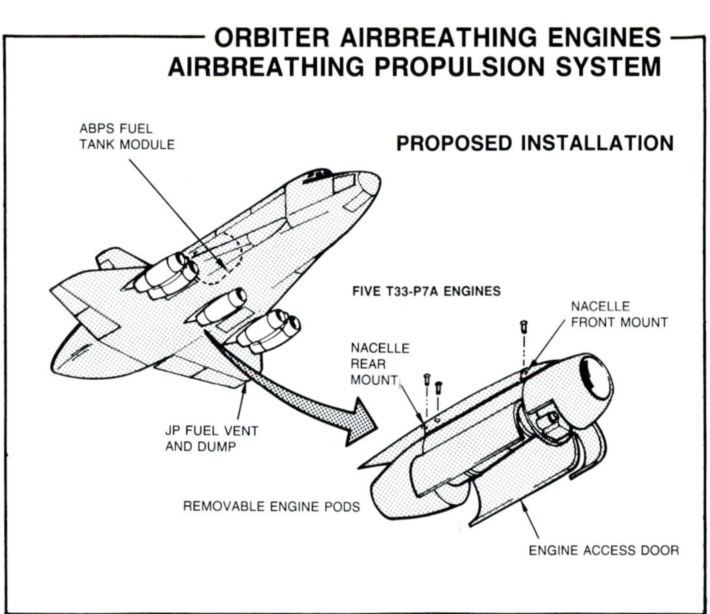

AUXILIARY POWER UNIT

WATER SPRAY UNIT

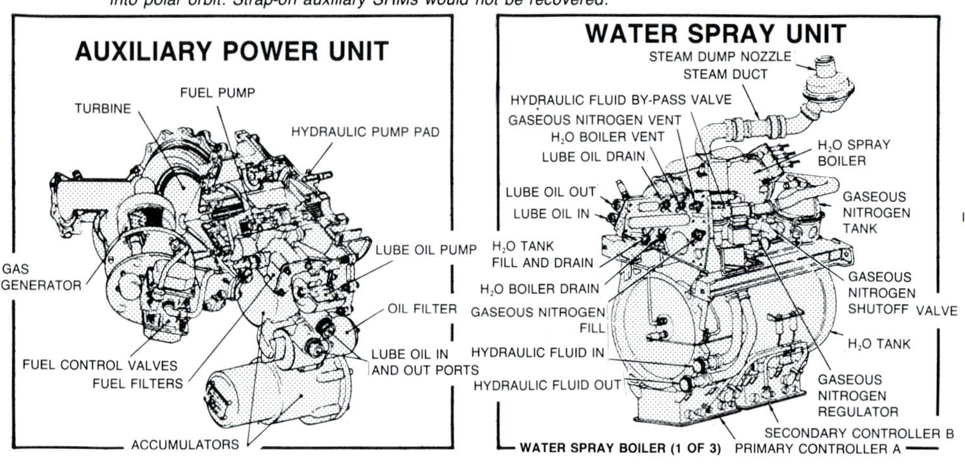

747 SHUTTLE CARRIER AIRCRAFT EMERGENCY ESCAPE SYSTEM

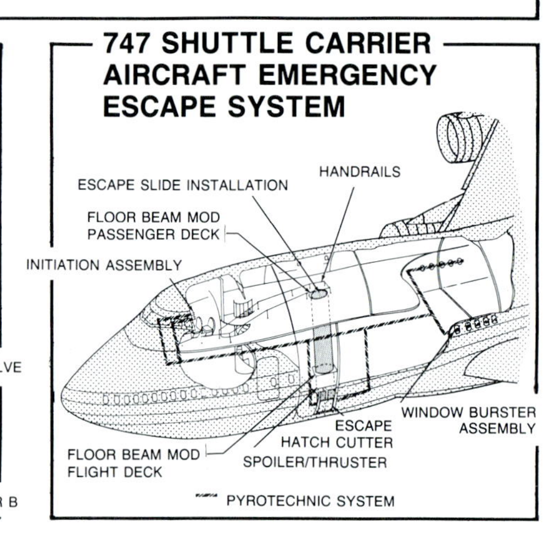

VENT SYSTEM

A. MAIN WHEEL WELL
B. AFT WING AND ELEVONS (SELF VENTING)
C. FORWARD WING AND WING GLOVE
D. ORBITAL MANEUVERING SYSTEM/REACTION CONTROL SYSTEM POD
E. FORWARD RCS CAVITY
F. NOSE WHEEL WELL
G. FORWARD FUSELAGE
H. PAYLOAD BAY & MID FUSELAGE
I. AFT FUSELAGE
J. VERTICAL TAIL
K. FORWARD WING AND WING GLOVE
L. MAIN WHEEL WELL
M. AFT WING AND ELEVONS (SELF VENTING)
N. OMS/RCS POD

L = LEFT VENT
R = RIGHT VENT

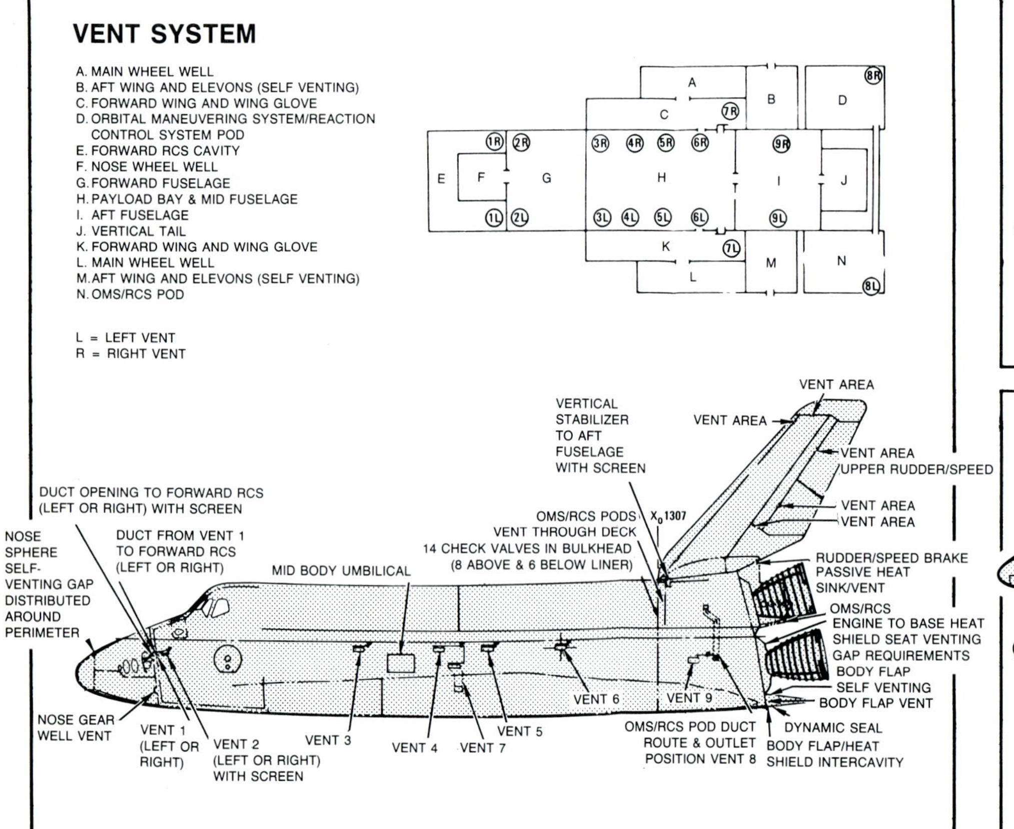

APU EXHAUST AND WATER BOILER VENT

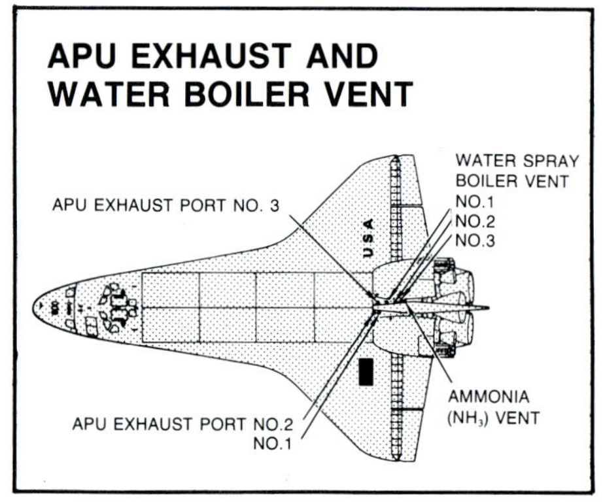

ORBITER VENT DOORS

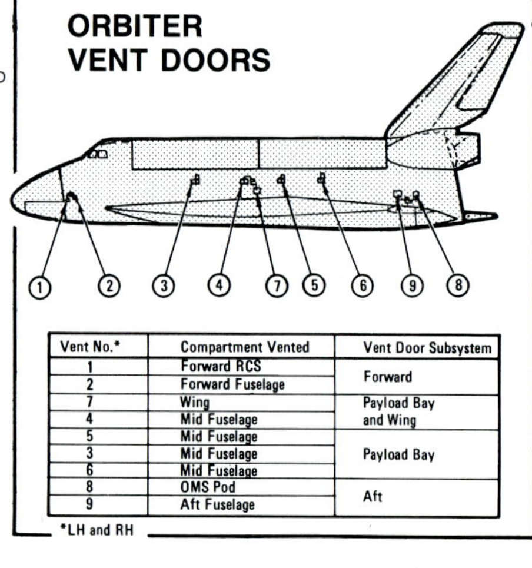

Vent No.*	Compartment Vented	Vent Door Subsystem
1	Forward RCS	Forward
2	Forward Fuselage	
7	Wing	Payload Bay and Wing
4	Mid Fuselage	
5	Mid Fuselage	Payload Bay
3	Mid Fuselage	
6	Mid Fuselage	
8	OMS Pod	Aft
9	Aft Fuselage	

*LH and RH

NASA

"Atlantis" at Pad B on October 9, 1986 in support of the launch pad check-out after STS-33/51-L. Tower houses 300,000 gals. of water for sound suppression system.

Dennis R. Jenkins

"Columbia" arrives at the Kennedy Space Center aboard the SCA. The attach points for the orbiter/SCA are the same as those for the orbiter/ET during launch.

Dennis R. Jenkins

A patch is visible on the nose cone of OV-101 where the pitot tube was fitted during preliminary flight tests.

USAF/AFFTC

Mate-demate device at DFRC is identical to the one at KSC. It provides work platforms for orbiter servicing, and also lifts it onto and off of the SCA. Simpler, transportable versions were placed at Vandenberg and White Sands.

NASA via Erik Simonsen

Mobile egress stairway permits quick crew off-loading following missions. "Columbia" is seen at Dryden following STS-4. This was the first landing on a hard runway (22).

NASA

Servicing convoy in place following "Challenger" KSC landing. Trucks provide power and cooling and inerting gases to the orbiter. A tow tractor already has been attached.

NASA

An astronaut stands on a workstand attached to the end of the remote manipulator arm during STS-11 (41-B). The enclosure for the WESTAR VI satellite is in the background.

NASA

Martin Marietta's MMU is the world's smallest manned spacecraft. Twenty-four nitrogen-powered thrusters provide 1.7 lbs. th. ea.; enough to reach 40 mph.

NASA

The two payload specialists for STS-25 (51-G) participate in a training exercise at Johnson Space Center. Noteworthy is the patch on Patrick Baudry which says "Challenger"—the flight actually was flown by "Discovery". Sultan Salman Abdul aziz Al-Saud (right) was Saudi Arabia's first astronaut.